PPM

JCB

CAT
Aveling Barford
HALL

BELL

FAUN
MOXY

O&K

GROVE CRANE
SAMSUNG

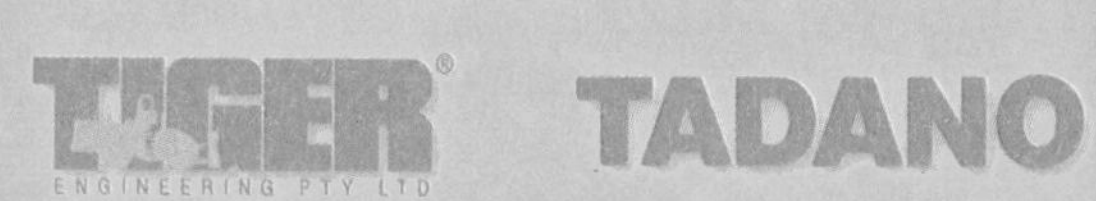
TIGER
TADANO

PPM

JCB

CAT

Aveling Barford
HALL

BELL

FAUN

MOXY

O&K

GROVE CRANE
SAMSUNG

TIGER
TADANO

PPM

JCB

GROVE

CAT

Aveling Barford

LIEBHERR
KOMATSU

BELL

FAUN

MOXY

O&K

GROVE CRANE

TIGER

PPM

JCB

CAT

Aveling Barford
HALL

BELL

KATO

KOBELCO

LIEBHERR
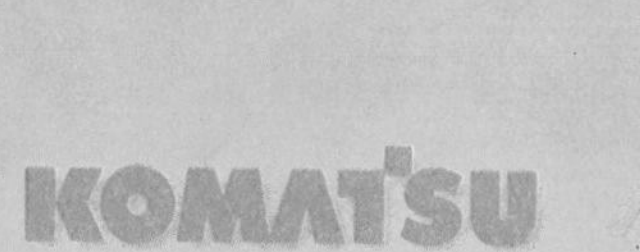
KOMATSU

# Heavy Equipment

# Heavy Equipment

## The World's Largest Machinery

**John Carroll**

A QUINTET BOOK

Published by Chartwell Books
A Division of Book Sales Inc.
114 Northfield Avenue
Edison, New Jersey 08837

This edition produced for sale in the U.S.A., its
territories and dependencies only.



ISBN 0-7858-0607-5

This book was designed and produced by
Quintet Publishing Limited
6 Blundell Street
London N7 9BH

Creative Director: Richard Dewing
Art Director: Silke Braun
Designer: Steve West
Project Editor: Clare Hubbard
Editor: Rosie Hankin
Picture Researcher: Penni Bickle

Typeset in Great Britain by
Central Southern Typesetters, Eastbourne
Manufactured in Bath, England
by DP Graphics
Printed in Singapore by
Star Standard Industries (Pte.) Ltd.

The author would like to acknowledge the help of
John Atherton of Alto Plant Services Ltd, Bardon (England) Ltd,
Ennemix Construction Materials, Tarmac Quarry Products
Eastern Ltd, and Ian Clegg.

Note

Tons (t) given in this book are US short tons. 1 ton = 2,000 pounds.
Metric tonnes (MT) are also provided. 1 metric ton = 2,204.6 pounds.

# Contents

# Introduction

**As long as there has been any form of civilization, human beings have sought through civil engineering to modify the environment around them to enable them to live more easily. As early as 510 BC, Darius, the King of Persia, ordered a canal to be cut from the River Nile to the Red Sea, the Romans built canals in various parts of Europe, and the Chinese built a series of waterways including the 600-mile (965.58-kilometer) long Grand Canal which in the eighth century is recorded as having carried 2.24 million tons (2.03 million tonnes) of goods. From medieval times onward, canal building spread widely across Europe but it was the Industrial Revolution that heralded the beginnings of the age of major civil engineering, of mechanization, and the development of the machine which enabled the scale of what was possible to increase exponentially. For example Ch'iao Wei-Yo, a Chinese engineer, is known to have invented the pound or chamber lock in order to lift and lower boats between different levels of waterway and had one built in AD 984. The principle was established and later Leonardo da Vinci would build a series of such locks on the Naviglio Interno near Milan in fifteenth-century Italy but it was the developments of the Industrial Revolution that enabled the construction of locks big enough to carry ocean-going ships. These developments affected both the undertaking and scale of civil engineering projects and the way in which they could be accomplished. Steam would provide the power for ships and railroad locomotives. It would also deliver the power for the machines used in their construction; it would provide the means to extract the fossil fuel for the machines. The machines would be capable of extracting minerals for the making of steel and of quarrying stone for construction.**

The history of machines for construction is intertwined with the economics of the labor market and economies of scale. The use of heavy machinery became widespread in the United States earlier than it did in Europe for reasons of economy: in the United States labor was scarce and expensive, so machines made economic sense, whereas in Europe labor was considerably more plentiful and cheaper. As a result much of the construction of the canal and, later, railroad systems in Europe was achieved through the efforts of countless laborers, known as "navvies" in Britain where many of their number were Irish. The Eastwick & Harrison Company from Philadelphia were amongst the pioneers of the mechanical excavator in the United States. Their limited-slew type Otis steam excavator was in use for railroad construction as early as 1838. By the 1930s the steam shovel was such an accepted part of American industry that mention of it by Woody Guthrie, the American Dust Bowl balladeer, would scarcely raise an eyebrow although his songs did: "...a dust storm buried her. She was a good girl; long, tall and stout. I had to get a steam shovel just to dig my darlin' out." ("Dust Bowl Blues").

# Canal Development

*One of the Ruston, Proctor & Co. Ltd Steam Navvies at work on the Manchester Ship Canal. Steam Navvies were named after the laborers who operated them.*

**Swift progress in engineering projects was made after the Industrial Revolution so that what was once only a dream soon became the achievable. An example is the Suez Canal that was designed as a short cut for ships between the Mediterranean and the Red Sea. The idea had been considered in as early as 1799 by Napoleon I of France but no progress was made for another six decades. Ferdinand de Lesseps was the man who provided much of the momentum behind the project and launched the International Suez Canal Company in 1858. The scale of the project was daunting, the line of the canal was 100 miles (161 kilometers) and it had to be wide and deep enough to carry ocean-going vessels. Despite British strategic concerns, work began in 1859 and was completed in November 1869.**

One of the last canals constructed in Great Britain was the Manchester Ship Canal, a 36-mile (57.93-kilometer) ditch between the sea and the inland city of Manchester. It was opened in 1894. The project required at various times up to 17,000 laborers and tradesmen, and temporary railroads were laid to carry wagons used for removing the soil. This massive project also provided a hint of what was to come as a great deal of

# Introduction

***One of the smaller steam excavators in the cut of the Manchester Ship Canal.***

machinery was also used. Steam-powered machines included 97 excavators, eight dredgers, 124 cranes, 192 portable and other engines, and 212 pumps. The excavators ran on rails including one manufactured by Ruston and Proctor of Lincoln to the patents of James Dunbar. This 35.84-ton (32.51-tonne) machine could excavate 2,000 cubic yards (1,529.20 cubic meters) in a 10-hour day. Similar smaller machines were manufactured by J.H.Wilson & Co. of Liverpool and Whitakers of Horsforth whose machines could excavate 400 and 500 cubic yards (305.84 and 382.30 cubic meters) per day. These machines soon became known as Steam Navvies after their human counterparts and established the practicality of heavy machinery for major works albeit some 50 years after their first use in the United States. Following this, developments came thick and fast from companies such as Bucyrus of Ohio and the Marion Steam Shovel Company, both of whom went on to become major manufacturers in the world of excavators.

The Panama Canal between the Atlantic and Pacific oceans is another example of a waterway built to carry ocean-going ships and of a massive civil engineering project. The finished waterway is only 40 miles (64.37 kilometers) long and yet saves ships a 6,000-mile (9,656-kilometer) trip around South America. Plans were initially formulated in the sixteenth century but a start was not made until late in the nineteenth by the same Ferdinand de Lesseps who had managed the construction of the Suez Canal. The Frenchman toiled for eight years on the project from 1879 amid allegations of fraud and mismanagement although he was hampered by the yellow fever and malaria which killed 22,000 workmen. The project foundered in 1889. Other routes for the canal were put forward and other companies considered the project, but the scheme was affected by political events such as a revolution in Panama. The United States made an agreement with Panama that, in return for some funds, a guarantee of Panamanian independence, and an annual payment, a strip of land for the canal could be found and subsequently administered by the United States. As well as advances in engineering being made, medical science was progressing to the extent that it was known that malaria and yellow fever could be controlled if the mosquitoes that carried the illnesses were eradicated. The decision was made to build a lock-type canal and work started in 1906. The project suffered some engineering setbacks but was completed under the supervision of Major G.W.Goethals, United States Army. The canal was opened immediately after the outbreak of the First World War. Numerous steam shovels were employed in the construction of the canal which, along with 45,000 workers, shifted 240 million cubic yards (183.50 million cubic meters) of earth. The cost of the project was $366,650,000.

***President Theodore Roosevelt observing the construction of the Panama Canal in November 1906.***

Canal construction projects such of these, and others such as The St. Lawrence Seaway in Canada, often require the cooperation of several nations and even after their completion can remain important within the political situation of a region. One of the United States' first acts after the outbreak of the Second World War in Europe was to increase the strength of its army through recruiting in order to reinforce its garrison in the Panama Canal zone. Similarly in the years after the Second World War the British and French armies occupied parts of Egypt in order to protect the Suez Canal. The operation was not wholly successful and the canal was subsequently blockaded for a number of years.

***A 1905 experiment with the 20 horsepower Hornsby oil engined tractor with a crawler conversion of 1896. It was the world's first fully tracked tractor as others used a tiller wheel to steer.***

# Tracked Machinery

**Paralleling the developments of the steam excavator were experiments with tracked machinery known as "crawlers." The first experiments involved wheeled steam tractors which were converted to run with tracks. The first test of such a machine took place in November 1904 in Stockton, California, where a Holt Steam tractor had been converted to run on tracks. This had been accomplished by the removal of the wheels and the rear's replacement with tracks made from a series of 3x4-inch (7.6x10-centimeter) wooden blocks bolted to a linked steel chain which ran around smaller wheels, a driven sprocket and idler on each side. Originally the machine was steered by a single tiller wheel although this system was later dropped in favor of the idea of disengaging drive to one track by means of a clutch which slewed the machine around. From there it was but a short step to gasoline-powered crawlers, one of which was constructed by Holt in 1906. By 1908 100 gasoline-powered crawlers were engaged in work on the Los Angeles Aqueduct project in the Tehachapi mountains.**

Necessity is generally considered to be the mother of invention and the two World Wars helped speed the development of heavy machinery in several ways. During the First World War the embryonic crawler technology was soon developed as the basis of the tank, now an almost universal weapon of war. During the Second World War the bulldozer earned numerous accolades and directly led to the blade-equipped tank, a type of armored fighting vehicle still in use by armies around the globe. American development and use of tanks lagged behind that of Europe partially because the United States remained uninvolved in the First World War until after the sinking of the Lusitania in 1917 through the German policy of unrestricted submarine warfare. Prior to this date the United States Army was still steeped in the cavalry traditions of the fighting in the Old West and by the time they became involved in the European conflict the European nations were using tanks on the western front. Orders went back to the United States for tanks but meantime American soldiers used British and French machines, namely the Mark VI and Renault FT17 respectively. These tanks were to be produced in the United States to take advantage of its massive industrial capacity although they had to compete for production line space with trucks and artillery so there was some delay. It is perhaps difficult to understand this when the crawler track was already well established in the United States and there had been a couple of experiments there too. The experimental machines included the Studebaker Supply Tank and the Ford 3.36-ton (3.05 tonne) tank. The French tank produced in the United States was designated the M1917 and was the only American tank-type to arrive in Europe prior to the armistice in 1918.

***This Sherman tank bulldozer was photographed in Normandy, France in July 1944. The armored dozer was used for a variety of purposes such as levelling obstructions and clearing roads.***

# Hydroelectric Engineering

*The Grand Coulee Dam in Washington State, the USA's largest concrete dam, was completed in 1941.*

Industrial development has occurred in phases as technology becomes practical which leads to various periods of history being remembered for different types of construction and civil engineering, examples being the canal age and the railroad age. More recently there have been huge hydroelectric schemes and massive road-building programs with the development of major roads referred to variously as freeways, motorways, and autobahnen. The United States engaged in several hydroelectric schemes in the late 1930s and early 1940s to provide electricity but also to provide work in order

**for the country to get itself out of The Great Depression. Two dams constructed at this time were the Hoover and Grand Coulee Dams. The Hoover or Boulder Dam is in Colorado at the Nevada/Arizona border. It stands 726 feet (221 meters) high—the highest concrete dam in America—and was completed in 1936. The Grand Coulee Dam across the River Coulee contains the Franklin D. Roosevelt lake which is 150 miles (241.40 kilometers) long. Situated west of Spokane, Washington, it is 550 feet (167.64 meters) high and 4,173 feet (1,271.93 meters) long—the USA's largest concrete dam—and was completed in 1942. Around the world there have been numerous other dams constructed as part of hydroelectric schemes; just two examples are the Kariba and Aswan High dams. The Kariba Dam was constructed in the late 1950s and is situated on the country that is now known as Zimbabwe (then Rhodesia) and Zambia. It dams the Zambesi River in the Kariba Gorge and was opened in 1960. It is operated by the governments of both countries although there were some problems during the final period of white majority rule in Rhodesia which led to a second generating station being built in Zambia. The Aswan High Dam was opened in 1971 and dams the River Nile in Egypt. Its purpose is to make electricity, and to prevent the Nile's floodwaters from being lost into the Mediterranean and so be available for irrigation of farmland.**

***Crawler technology has progressed rapidly since World War Two and the size and capability of the machines has increased dramatically. This O&K RH120-C weighs in excess of 110.23 tons (100 tonnes).***

The development of the machinery for use in massive construction and mineral extraction has become increasingly sophisticated over the years. The initial cost of purchasing such machinery means that much of it must have a viable working life measured in years and that it must be used at its optimum level. This will determine which machinery is best suited for a specific task. A wheeled loader, for example, can only be efficiently operated if the dump trucks being loaded can remove the rock at the same rate at which it can be loaded otherwise the loader's capability is being wasted. In industries dependent on the price of raw materials this, along with other factors, is of paramount importance. The remainder of this book will consider this and other factors in describing a representative and comprehensive selection of the world's heavy equipment.

950E

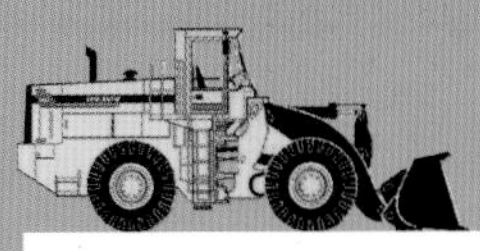

# Wheeled Loaders

# Wheeled Loaders

**The loader, be it wheeled, tracked, or a backhoe, is an established machine in the various tasks that can be loosely grouped within earthmoving and vital to many of them, specially in strip mining and quarrying where large quantities of minerals need to be transported from where they are excavated. The loader is also suitable for specialist application such as loading felled trees for the timber industry. In order to make loaders suitable for as many applications as possible, most manufacturers offer a variety of specialized components particularly in the size and type of bucket fitted and, in many cases, special high lift options are available. The buckets are susceptible to damage from being continually exposed to broken rock and other abrasive materials so are increasingly being designed with replaceable parts to allow for wear and tear.**

Large wheeled loaders are manufactured by Atlas, Case, Caterpillar, Daewoo, Halla, Fiat-Hitachi, Hyundai, JCB, Kawasaki, Komatsu, Liebherr, O&K, Samsung, and Volvo BM. Tracked loaders are produced by Caterpillar, Fiat-Hitachi, and Liebherr. In this chapter, as with subsequent ones, a representative sample of machines is considered in detail.

*The wheeled loader such as this Caterpillar 950E machine is ideally suited to mineral extraction through its combination of maneuverability and lifting capability.*

# Caterpillar 980F Series II Wheeled Loader

**The Caterpillar 980F Series II wheeled loader is powered by Caterpillar's own six-cylinder 3406C turbocharged four-stroke cycle direct-injection engine. The engine delivers full rated net power of 205 kilowatt (275 horsepower) and has a high torque rise for superior lugging. This latter feature ensures that the machine and material can be kept moving with a minimum of downshifting. The four-stroke cycle delivers a long power stroke for improved fuel combustion. The whole unit is designed so that it is able to deliver reliable, effective performance at the minimum cost and with only basic servicing.**

The engine turbocharger offers increased fuel economy and improved high-altitude performance. It is aftercooled for clean efficient combustion of fuel and reduces heat build-up. The high pressure direct-injection fuel system injects fuel at up to 103 350 kilometers per acre for precise metering. The fuel is fully atomized which allows almost total combustion; it is an adjustment-free system which keeps maintenance required to a minimum. The 3406C turbocharged four-stroke cycle direct-injection engine features full-length watercooled cylinder liners designed to dissipate heat evenly so reducing piston, piston ring, liner, and block wear. The pistons are oil cooled for similar reasons of longevity. The oil is cooled through use of an oil cooler designed to cool the oil that is in turn cooling the engine components. The remainder of the cooling is carried out by a multi-row modular radiator which fulfills the dual objectives of cooling and noise reduction. A sight gauge is fitted that allows the operator to check on coolant levels and, a swing-out grille gives easy access. Equally accessible for maintenance purposes are the engine oil dipstick, air cleaners, alternator, fuel filters, engine oil filters, and air-conditioner compressor.

***The Cat 980F is powerful and easy to maneuver, allowing for fast, efficient output.***

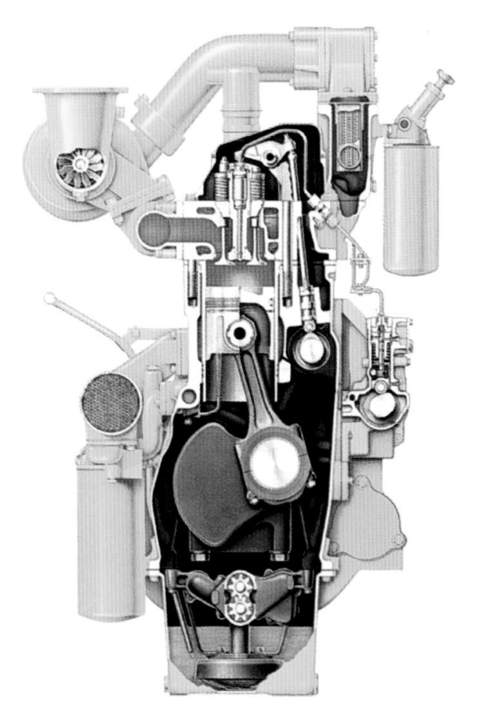

***A cross section of the Caterpillar 3406C engine. It is of a four stroke cycle type with a long power stroke for maximum fuel combustion and performs at a full-rated net power of 205 kilowatts (275 horsepower).***

Backing up this complex power plant is Cat's field-proven planetary transmission. This provides automatic shifting for simple reliable operation. Clutches are designed to ease together during shifts to reduce shocks to the power train. The transmission is designed with perimeter-mounted, large-diameter clutch packs that easily control inertial loads in the transmission in order to prolong component life. The transmission features a large gear contact area, and the heat-treated clutch plates are oil cooled to enhance longevity and increase time between services. The transmission is electronically controlled and shifts at factory-set intervals. This preset system ensures a precise match of engine torque to ground speed and has the benefit of fewer moving parts.

The all-wheel drive wheeled loader's axles are designed for heavy duty applications, long service, and to cope with severe conditions. The final drive planet gears and differential pinions use full-floating bronze sleeve bearings. A seal patented by Caterpillar known as the Duo-Cone is fitted between the axle shafts and the housings in order to seal the lubrication in and the dirt out. The axles are designed to keep all wheels on the ground for stable operation even in arduous terrain. The differentials fitted as standard are of the limited slip type although there is an optional rear differential known as the NoSPIN. The purpose of using limited slip differentials is to ensure maximum traction is retained even in the wettest and softest of conditions. The final drives feature planetary reduction at each wheel and can be removed independently of the wheels and brakes from the machine. The 980F is fitted with an oil-cooled, multiple-disc braking system which is fully sealed and adjustment free. It uses large-diameter discs and plates in order to provide fade-resistant braking capability, and the oil runs in face grooves to eliminate heat fade. This system is backed up by a parking brake which can be engaged by the operator but also automatically engages if brake pressure drops as a fail-safe system.

The loader's structures are built to a sufficient quality to enable them to work on the hardest applications in

***The front section of the 980F is constructed around a full box section frame (1) which maximizes the distance between the upper and lower hitch to minimize twisting forces (2). The loader tower (3) is of a four plate design to handle severe loading stresses and provide mounts for the solid steel lift arms (4). These latter components are connected by an oval section cast steel cross tube (5).***

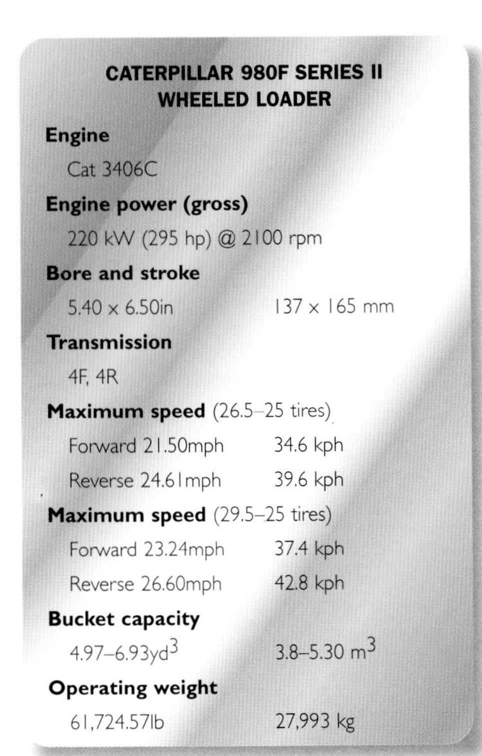

| CATERPILLAR 980F SERIES II WHEELED LOADER | |
|---|---|
| **Engine** | |
| Cat 3406C | |
| **Engine power (gross)** | |
| 220 kW (295 hp) @ 2100 rpm | |
| **Bore and stroke** | |
| 5.40 x 6.50in | 137 x 165 mm |
| **Transmission** | |
| 4F, 4R | |
| **Maximum speed** (26.5–25 tires) | |
| Forward 21.50mph | 34.6 kph |
| Reverse 24.61mph | 39.6 kph |
| **Maximum speed** (29.5–25 tires) | |
| Forward 23.24mph | 37.4 kph |
| Reverse 26.60mph | 42.8 kph |
| **Bucket capacity** | |
| 4.97–6.93yd$^3$ | 3.8–5.30 m$^3$ |
| **Operating weight** | |
| 61,724.57lb | 27,993 kg |

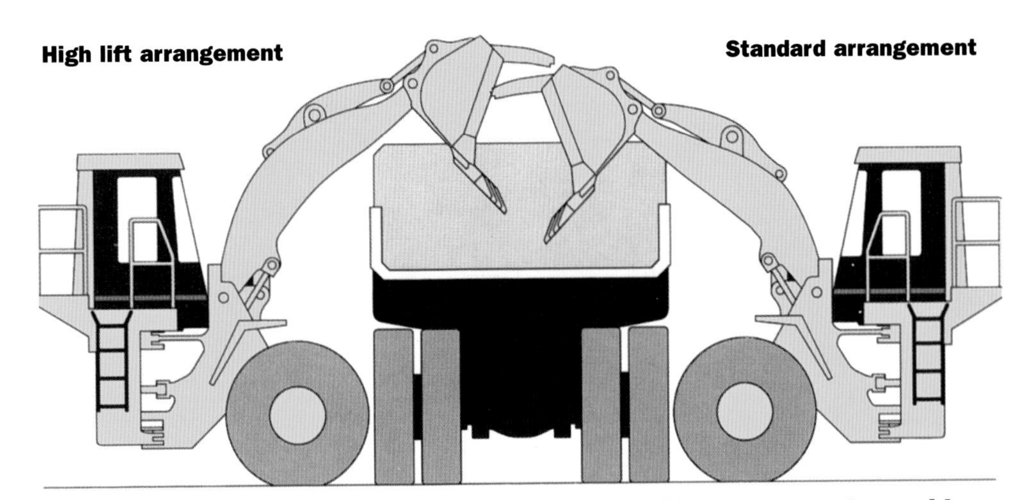


***Different length lift arms are available; the high lift arrangement provides higher dump clearance and longer reach, but the standard arrangement has a higher lift capacity.***

the toughest environments. Construction is based around a full box section frame which is resistant to twisting and impact forces in order to keep the power train components solidly aligned. The spread hitch design widens the distance between upper and lower hitches to avoid twisting in order to maximize hitch pin and roller bearing life. Castings are used in high-stress areas to add strength and diffuse stress loads. The loader tower is of a four-plate design, built to handle loading stresses and create a solid mounting for the lift arms while shielding the hydraulic system components from impacts and debris. The lift arms are made from solid steel plate and are connected by an oval section cast steel cross tube which is tubular to prevent twisting and flexing, thus maintaining pin bore alignment. The hitch points–upper and lower–are mounted on tapered roller bearings to distribute loads evenly. The loader linkage is of a Z-bar design which has as few pivot points and moving parts as possible to reduce servicing time and costs. The linkages are sealed in order to retain pin lubrication for as long as possible, and the Z-bar system is engineered to maximize breakout forces as it is these forces that ensure a full bucket load. It is of course the hydraulic systems that make the whole loader viable and Cat's designers have maximized breakout forces, bucket loading times, and steering control through judicious use of hydraulics. The lift and tilt cylinders are of a large bore diameter to gather and lift speedily. The bucket controls are operated with finger touch from the cab and give precise control over the loading functions. The steering system is designed to ensure that the machine is easy to maneuver in all ground conditions. The machine's hoses are assembled using o-ring face seals and replaceable couplings.

Because of the specific nature of some loading tasks, the 980F has a number of options that it can be supplied with. In place of the bucket a log loader attachment can be fitted. This jawlike arrangement, known as a grapple fork, has widely spaced tines to enable tree-length logs to be carried and loaded. For specialist applications with a loading bucket, a high lift arrangement can be supplied which increases the dump clearance and longer reach of the standard 980F Series II. With this option the machine is suitable for loading trucks as large as the 39.20-ton (35.56-tonne) off-highway trucks. The 980F can be supplied with a variety of buckets and ground engaging tools. These include both general purpose and rock buckets.

***According to Cat the 994 is the world's largest drive loading tool.***

# Caterpillar 994 Wheeled Loader

***The design of the lifting arms of the Caterpillar 994 is intended to maximize visibility from the cab for both excavating (above) and loading (below) operations through the use of low profile arms and tilt cylinders.***

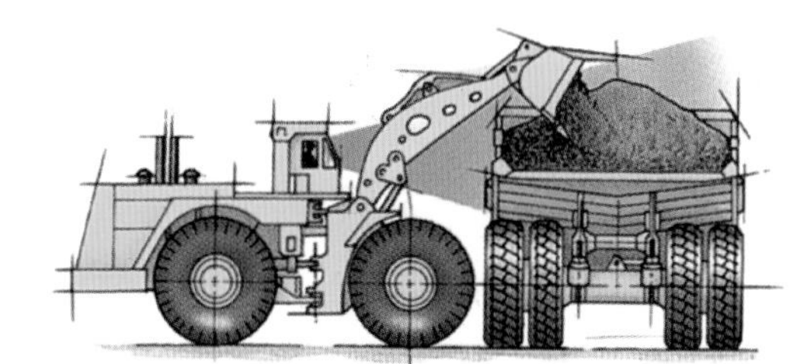

**The 994 is a wheeled loader also built by Caterpillar but is considerably larger than the 980F Series II. The 994 is, according to its manufacturer, the world's largest mechanical drive loading tool. It is designed to load trucks rated at 168.00 tons (152.41 tonnes) and up through a bucket payload of between 33.60 and 39.20 tons (30.48 and 35.56 tonnes). The 994 is designed to work in conjunction with 168.00-ton (152.41-tonne) trucks which it loads to capacity in four passes, and correspondingly more passes for larger capacity trucks. It is the first product from the Caterpillar Mining Vehicle Center, a facility for building mining vehicles, and is intended as a primary production loading tool or a mobile loader to back up face shovels.**

In some ways it is a bigger version of the smaller wheeled loaders in that it features the Z-bar loader linkage, four-plate design loader arms, and solid plate lift arms. The 994 uses the proven Cat 3516 turbocharged diesel engine of a four-stroke cycle design with a direct-injection fuel system. The engine block is a single casting strengthened with internal ribbing. The engine features four turbochargers and a water circuit after cooler. Each cylinder has its own separate cylinder head. The valve configuration is of two inlet and two exhaust valves per cylinder, the valves are hard faced, and the valve seats are hard alloy steel. The pistons are two piece, ferrous crowned with aluminum skirts and three piston rings which are cooled by oil spray. The engine, which is estimated to last 15,000 hours before needing a rebuild, is cooled by

**CAT 994 WHEELED LOADER**

| | | |
|---|---|---|
| **Engine** | | |
| Cat 3516 | | |
| **Engine power (gross)** | | |
| 996 kW (1336 hp) @ 1600 rpm | | |
| **Bore and stroke** | | |
| 6.70 x 7.49in | | 170 x 190 mm |
| **Transmission** | | |
| 3F, 3R | | |
| **Maximum speed** (49.5-57 L-4 tires) | | |
| Forward | 14.17mph | 22.8 kph |
| Reverse | 15.22mph | 24.5 kph |
| **Maximum speed** (53.5/85 L-5 tires) | | |
| Forward | 15.35mph | 24.7 kph |
| Reverse | 16.53mph | 26.6 kph |
| **Bucket capacity** | | |
| 13.08–40.55yd3 | | 10-31 m3 |
| **Operating weight** | | |
| 385,434lb | | 174,800 Kg |

a multi-row modular radiator. Its transmission is a larger version of Cat's planetary power shift transmission. The major way that it differs from the 980F is in size; the 994 almost dwarfs the 980F.

Its massive loader-dozer design tires are mounted on demountable rims which are mounted on the pair of driven axles. The front axle is fixed, although the rear one can rise or drop 26.70 inches (678.18 millimeters), with all wheels remaining in contact with the ground for maximum traction. The operator's cab has standard safety features such as rollover protection systems, resiliently mounted cab to dampen vibration, and a suspension equipped seat. Air conditioning and filtration are also fitted. The visibility for the operator is maximized through the positioning of the operator high in the cab which is set as far forward as possible. The designers realize that machine productivity is dependent on operator productivity, which is influenced by factors such as ease of control, comfort, visibility, and confidence in the safety of the machine.

Another factor that affects productivity is the match of loader to truck fleet. A poorly matched system with insufficient loader capacity leads to trucks waiting to be loaded, a waste of man and machine hours. To maximize this potential, an optional high lift arrangement is available to enable the loader to load higher and larger trucks.

***Optimum productivity of the wheeled loader is dependent on the operator. Caterpillar have designed the operator's seat with suspension to enable controls to be easily reached.***

# Halla Wheeled Loaders

***The Halla HA380 Wheeled Loader has a minimal turning radius because of its 45 degrees steering angle, which makes for easier operation in confined spaces. Steering is accomplished by means of a hydraulic turning system.***

**Halla Engineering and Heavy Industries Limited are based in Seoul, Korea, and manufacture a range of wheeled loaders including the HA170, HA290 and HA380, the HA380 being the largest.**

The machines are modern and feature a number of proprietary components including Cummins engines, and German-made torque converters and transmission. The Cummins unit fitted is a six-cylinder, naturally aspirated, direct-injection, four-cycle diesel. The transmission is a full-power shift countershaft design with four forward and three reverse speeds. It is controlled by a single lever on the steering column. Steering is hydraulic and a load-sensing priority valve is

**HALLA HA380 WHEELED LOADER**

**Engine**
Cummins LTA 10-C

**Engine power (flywheel)**
216 kW (287 hp)

**Bore and stroke**
4.93 x 5.36in 125 x 136 mm

**Transmission**
4F, 3R

**Maximum speed** (26.5x25 (20PR) tires)
Forward 14.29mph 37 kph
Reverse 10.35mph 26.8 kph

**Bucket capacity**
4.97yd$^3$ 3.8 m$^3$

**Operating weight**
47,539.80lb 21,560 kg

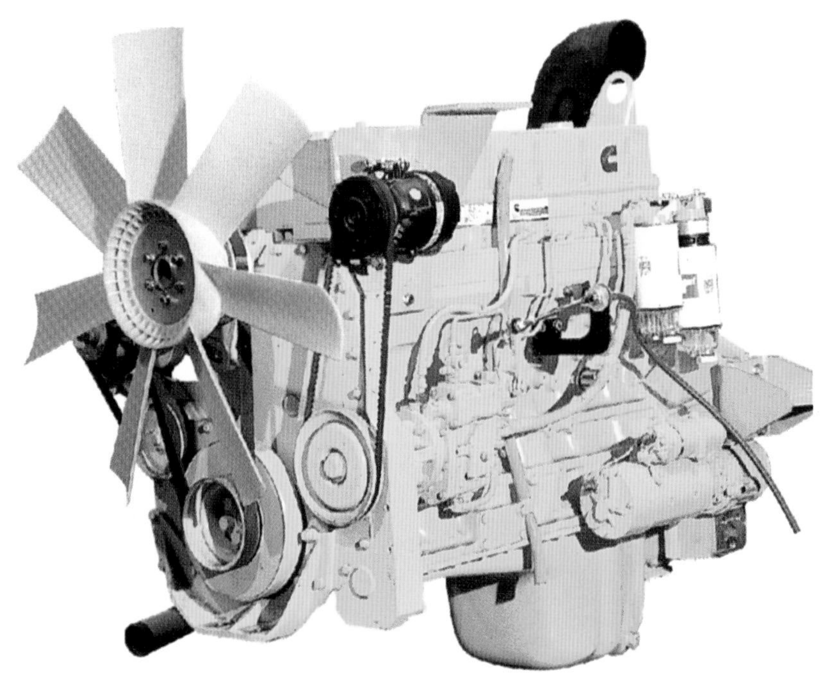

***The Halla HA380 is powered by a 216 kilowatt (287 horsepower) turbo-charged Cummins direct-injection six-cylinder diesel engine of 611 cubic inches (10,014 cubic centimeter) displacement.***

***Operation of the HA380 in severe terrain is made possible by the oscillation of the rear axle which is mounted by a center pin. The pin allows movement of up to 24 degrees which equates to 18.3 inches (470 millimeters) of axle travel.***

fitted. The axles have planetary reduction gears at the wheel ends and of these the front one is fixed while the rear has 24 degrees of oscillation which equates to 18.52 inches (470 millimeters) of vertical travel.

*The L15 has a bucket capacity of 1.96 cubic yards (1.5 cubic meters).*

# O&K Wheeled Loaders

*The L35 has formidable off road capabilities.*

**O&K–Orenstein and Koppel AG–are based in Dortmund, Germany, and manufacture a comprehensive range of 12-wheeled loaders. The capacity of the buckets on these machines ranges from 0.59 cubic yards (0.45 cubic meters) to 4.97 cubic yards (3.8 cubic meters) in the L4B and L45B machines respectively. The L15 has a bucket capacity of 1.96 cubic yards (1.5 cubic meters) and an operating weight of 18,929.93 pounds (8,585 kilograms), the L25 of 3.01 cubic yards (2.3 cubic meters) and 27,694 pounds (12,560 kilograms)**

**O&K L45B WHEELED LOADER**

**Engine**
LT-10-C 250

**Engine power (gross)**
177 kW (240 hp) @ 2100 rpm

**Bore and stroke**
n/a

**Transmission**
4F, 4R

**Maximum speed** (26.5–R25 tires)
Forward 23.18mph 37.3 kph
Reverse 23.18mph 37.3 kph

**Bucket capacity**
4.97yd$^3$ 3.8 m$^3$

**Operating weight**
49,149.45lb 22,290 kg

***The O&K L45 is one of the largest machines in the company's range of wheeled loaders and is powered by a 177 kilowatt (240 horsepower) Cummins diesel engine. This makes the machine capable of a maximum speed of 23.18 miles per hour (37.3 kilometers per hour.)***

**respectively, while the L35B holds 4.19 cubic yards (3.2 cubic meters) and weighs 40,263 pounds (18,260 kilograms).**

O&K have designed their small and mid-sized crawler mounted excavators, wheeled excavators, and wheeled loaders to be economic performers on site. In 1986 the company launched an electronic three pump control for its range of hydraulic excavators. The hydraulic technology was also applied to O&K's range of wheeled loaders with shovel capacities ranging from 0.78 to 4.97 cubic yards (0.6 to 3.8 cubic meters). The machines have low noise emission standards and are both strong and maneuverable. They are designed to offer high loading performance with low fuel consumption for economic operation.

***While this L25 has a simple bucket for moving earth, reinforced buckets are used in rock shifting situations.***

# Samsung Wheel Loader Series-2

**SAMSUNG SL150-2 WHEELED LOADER**

| | | |
|---|---|---|
| **Engine** | | |
| | Cummins 6BT5.9-C | |
| **Engine power (flywheel)** | | |
| | 86 kW (116 hp) @ 2400 rpm | |
| **Bore and stroke** | | |
| | 4.02 x 4.73in | 102 x 120 mm |
| **Transmission** | | |
| | 4F, 3R | |
| **Maximum speed** (17.5x25 tires) | | |
| | Forward 21.25mph | 34.2 kph |
| | Reverse 12.80mph | 20.6 kph |
| **Bucket capacity** | | |
| | 2.22yd$^3$ | 1.7 m$^3$ |
| **Operating weight** | | |
| | 22,535.10lb | 10,220 kg |

**SAMSUNG SL250-2 WHEELED LOADER**

| | | |
|---|---|---|
| **Engine** | | |
| | Cummins LTA10-C | |
| **Engine power (flywheel)** | | |
| | 191 kW (256 hp) @ 2100 rpm | |
| **Bore and stroke** | | |
| | 4.93 x 5.36in | 125 x 136 mm |
| **Transmission** | | |
| | 4F, 3R | |
| **Maximum speed** (23.5x25 tires) | | |
| | Forward 23.12mph | 37.2 kph |
| | Reverse 16.22mph | 26.1 kph |
| **Maximum speed** (26.5x25 tires, optional) | | |
| | Forward 25.29mph | 40.7 kph |
| | Reverse 17.65mph | 28.4 kph |
| **Bucket capacity** | | |
| | 4.58yd$^3$ | 3.5 m$^3$ |
| **Operating weight** | | |
| | 46,040.40lb | 20,880 kg |

**Samsung Heavy Industries are based in Korea and produce the SL Series of wheel loaders which are designed with both performance and versatility in mind. The machines are up to date in every respect and designed for a range of bucket payloads. The four SL Series machines range from a heaped bucket capacity of 2.22 cubic yards to 4.58 cubic yards (1.7 cubic meters to 3.5 cubic meters). The loaders feature axles equipped with torque proportioning differentials–limited slip–which improve traction and tire life. The rear axle pivots to improve traction further on rough terrain, and the axles are designed so that they can be partially dismantled for servicing without removing wheels, tires, and planetary reduction components in the hubs. The wheeled loader has 45 degrees of articulation to allow maximum maneuverability and steering is power assisted. The braking system incorporates wet discs; these are fully enclosed on each wheel to ensure maximum brake life. The four-speed transmission is described by Samsung as "Powershift" and controlled by a twistgrip lever mounted on the steering column. It includes features such as an automatic kick down switch.**

The SL Series wheeled loaders are powered by six-cylinder Cummins diesel engines, ranging from a 358 cubic inch (5,866.55 cubic centimeter) model in the SL150-2 to a 610 cubic inch (9,996.07 cubic centimeter) unit in the SL250-2. They are of the watercooled, direct-injection design.

The SL Series have maximized dump clearance (the distance between the ground and the lower edge of the bucket when it is positioned to deposit its contents) and reach (the distance over a truck that the bucket can reach to deposit its load). The bucket is operated by Samsung's design of Z-bar linkage.

The rollover protected cab is designed to make operation as easy as possible; a pillarless front cab window maximizes visibility for the operator, while a tiltable steering wheel and fully adjustable Bostrom seat allows room for both large and small operators. The cab is rubber mounted to minimize the vibration transmitted to the operator, and is fully air conditioned and pressurized. The functions of the entire machine are controlled from within the cab; a self-diagnostic system highlights and identifies system faults through a panel display and a one-touch control panel is fitted for a variety of the functions.

***Because of their lifting capacity and maneuverability, wheeled loaders are popular for quarrying applications.***

# Volvo BM L-Series

*A Volvo L180C being used for the extraction of mineral sand. The L180C is one of the mid-range wheeled loaders manufactured by Volvo. A variety of bucket sizes and a long boom option can be specified depending on the use to which the loader will be put.*

**The current range of wheeled loaders made by Swedish auto giant Volvo is the C-Generation, of which there are seven models of varying capability. The L50C is the smallest of the range, while the L330C is the largest. In between are the L70C, L90C, L120C, L150C, and L180C models. The machines are designed for optimum rock handling with adequate breakout force and durability required for quarrying purposes. The front frame is ruggedly designed with cast attachment points, fully floating axles, and fully enclosed disc brakes. The engine and drivetrain combination is Volvo designed and manufactured. The transmission is of the automatic type tagged APS II by the manufacturer. The L50 to L160 models feature what the manufacturers term CDC–Comfort Drive Control. This eliminates the monotonous and strenuous turning of the steering wheel right and left in repeated operating cycles, reducing strain on the operator. With CDC the operator can steer and shift gear with the left hand and operate the loader linkage with the right. The operating lever is fitted to a collapsible left arm rest alongside the seat, and steering speed is proportional to the stroke of the lever making it possible to inch the loader.**

**VOLVO L50C WHEELED LOADER**

**Engine**
TD 40 KAE

**Engine power (flywheel)**
71 kW (97 hp) @ 2200 rpm

**Bore and stroke**
n/a

**Transmission**
2F, 2R

**Maximum speed** (15.5R25 tires)

| | |
|---|---|
| Forward 24.23mph | 39 kph |
| Reverse 24.23mph | 39 kph |

**Bucket capacity**

| | |
|---|---|
| $1.83yd^3$ | $1.4\ m^3$ |

**Operating weight**

| | |
|---|---|
| 18,257.40lb | 8,280 kg |

A Volvo L50C being used for gravel extraction. Economic operation of such machinery depends on its compatability with the size of truck to be loaded.

# VOLVO

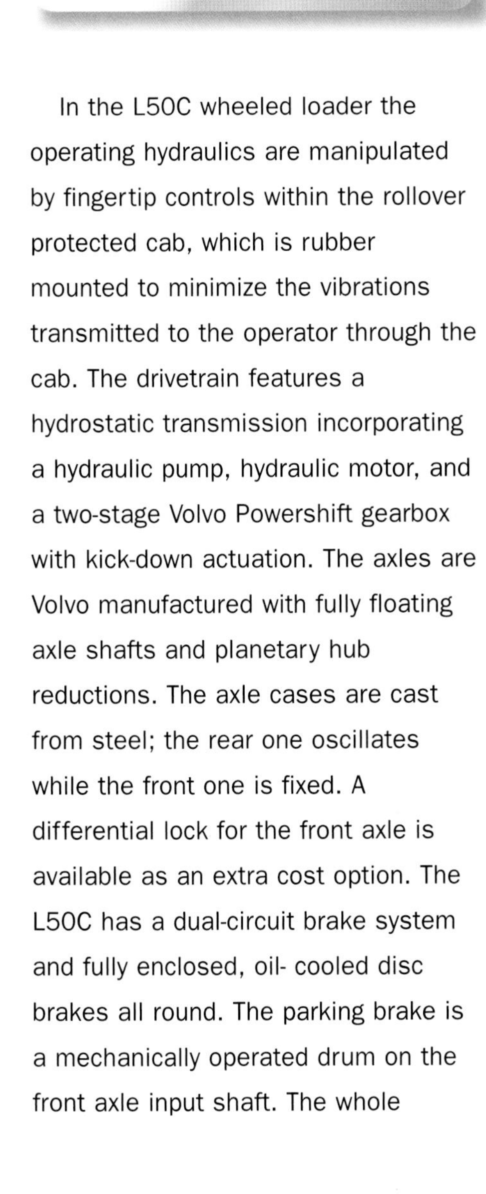

In the L50C wheeled loader the operating hydraulics are manipulated by fingertip controls within the rollover protected cab, which is rubber mounted to minimize the vibrations transmitted to the operator through the cab. The drivetrain features a hydrostatic transmission incorporating a hydraulic pump, hydraulic motor, and a two-stage Volvo Powershift gearbox with kick-down actuation. The axles are Volvo manufactured with fully floating axle shafts and planetary hub reductions. The axle cases are cast from steel; the rear one oscillates while the front one is fixed. A differential lock for the front axle is available as an extra cost option. The L50C has a dual-circuit brake system and fully enclosed, oil- cooled disc brakes all round. The parking brake is a mechanically operated drum on the front axle input shaft. The whole machine is powered by an in-line, four-cylinder, direct-injection, turbocharged diesel engine. The steering is by means of a load-sensing hydrostatic articulated system, as is the hydraulic system for the remainder of the loader's functions.

***The side elevation of the smallest Volvo wheeled loader, the L50C. It has a wheelbase of 108.2 inches (2,750 millimeters) and an overall length to the rear of the bucket in this position of 205.5 inches (5,220 millimeters).***

***The side elevation of the largest Volvo wheeled loader, the L330C. It has a wheelbase of 159.8 inches (4,060 millimeters) and an overall length to the rear of the bucket in this position of 342.5 inches (8,700 millimeters.) Overall height is 164.1 inches (4,170 millimeters).***

**VOLVO L330C WHEELED LOADER**

**Engine**
Volvo TD 164 KAE

**Engine power (flywheel)**
370 kW (503 hp) @ 1800 rpm

**Bore and stroke**
n/a

**Transmission**
4F, 3R

**Maximum speed** (35.65R33 tires)

| | | |
|---|---|---|
| Forward | 21.25mph | 34.2 kph |
| Reverse | 12.37mph | 19.9 kph |

**Bucket capacity**
8.63yd$^3$ 6.6 m$^3$

**Operating weight**
105,641.55lb 47,910 kg

The Volvo L330C is the largest machine in Volvo's range of wheeled loaders and, although similar in many respects, by dint of its design needs to incorporate more powerful major components such as engine and transmission. It is designed for the primary loading of 71.65-ton (65-tonne) haulers in quarries and mines.

The L330C is powered by an in-line, six-cylinder, direct-injection, turbocharged, intercooled diesel engine of a low emission design which is concealed behind large engine access doors. The transmission features a single stage torque converter, a power shift, countershaft with single lever control, and Volvo's APS II shift system with mode selector. The axles are based on cast steel cases with planetary type hub reductions, of which the front one is fixed and the rear oscillates to allow all wheels to remain in contact with the ground in uneven conditions. Posi-Torq limited slip differentials are fitted to both axles. Braking is by means of outboard-mounted, oil-cooled wet discs at each wheel. Transmission declutch when braking can be preselected by a switch on the instrument panel. The steering is power assisted and load sensing, and is powered by two variable flow axial piston pumps. The hydraulics for the bucket are similar and powered by four load-sensing, variable-flow axial piston pumps. The bucket is operated via a Z-bar linkage proven for loaders because of its breakout capabilities. The cab is air conditioned, vibration suppressed, and features a suspended operator's seat. As well as this, good all-round visibility is ensured through use of a curved tinted windshield. There is a proven link between operator comfort and productivity.

***The wheeled loader is at its most useful when moving loose rock whether it is to be loaded into a crusher or a truck.***

# Specialist Loaders

***A variety of specialized tools for rock handling are available including pallet forks and buckets, all are capable of being mounted to the 988F.***

**There are a variety of variations on the theme of wheeled loaders, of which the most extreme is the tracked loader. However there are wheeled loaders fitted with equipment to enable them to carry out specific tasks. Examples include block, timber, and waste handlers.**

Caterpillar are one company who manufacture block handlers, based on their 988F Series II wheeled loader. There are two lift arrangements, for heavy and standard applications. The heavy system ensures optimum machine stability and lift capacity for handling large blocks through the use of shorter heavy duty lift arms, larger tilt cylinders, high pressure hydraulic system, a counterweight, and special tilt links. The standard machine is intended to provide more versatility when a combination of forks and buckets is being used for handling blocks. The standard system is the same as the heavy system with the exception of the length of the lift arms. Specific block-handling tools are manufactured and can be used in conjunction with the loader: a block-handling pallet fork, quick coupler, breaker tine, and clearing rake.

Caterpillar also offer log handlers based on the 966F, 980G, and 988F Series II models. Another of the companies who manufacture log handlers are Volvo BM. They offer log-handling tools on their L50C, L70C, L90C, L120C, L150C, and L180C machines. Eight specialized grapples are available for the machines depending on the nature of the specific task to be carried out. The cab, chassis, and mechanical details are the same as the standard wheeled loaders while the payload of the grapple increases with the designation. The largest offered is the 19,404 pounds (8,800 kilograms) lift available on the L180C machine.

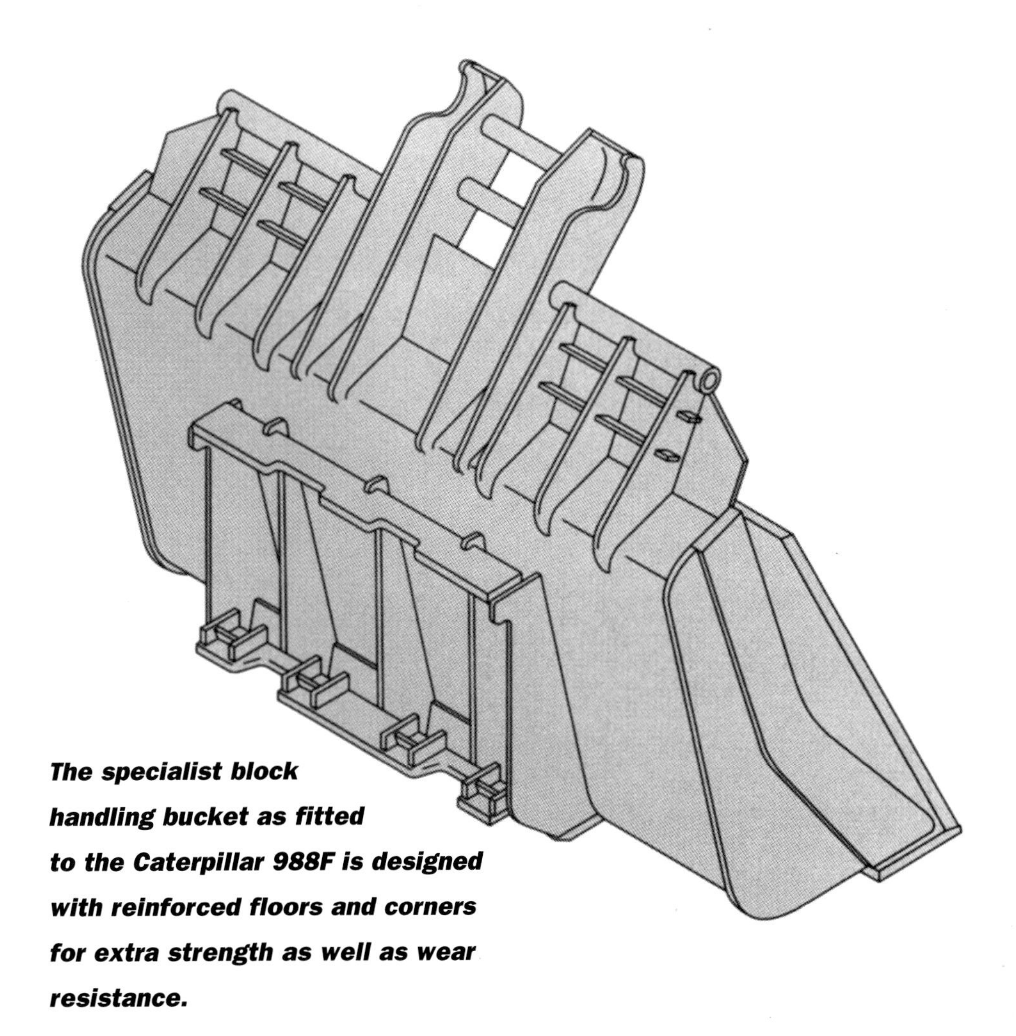

***The specialist block handling bucket as fitted to the Caterpillar 988F is designed with reinforced floors and corners for extra strength as well as wear resistance.***

**CATERPILLAR 988F SERIES II BLOCK HANDLER**

**Engine**
Cat 3408E

**Engine power (gross)**
342 kW (458 hp) @ 2000 rpm

**Bore and stroke**
5.40 x 5.99in — 137 x 152 mm

**Transmission**
4F, 3R

**Maximum speed** (35/65-33 24 PR (L-4) tires)
Forward 21.75mph — 35 kph
Reverse 14.60mph — 23.5 kph

**Operating weight**
118,653.26lb — 53,811 kg
(variable depending on options fitted)

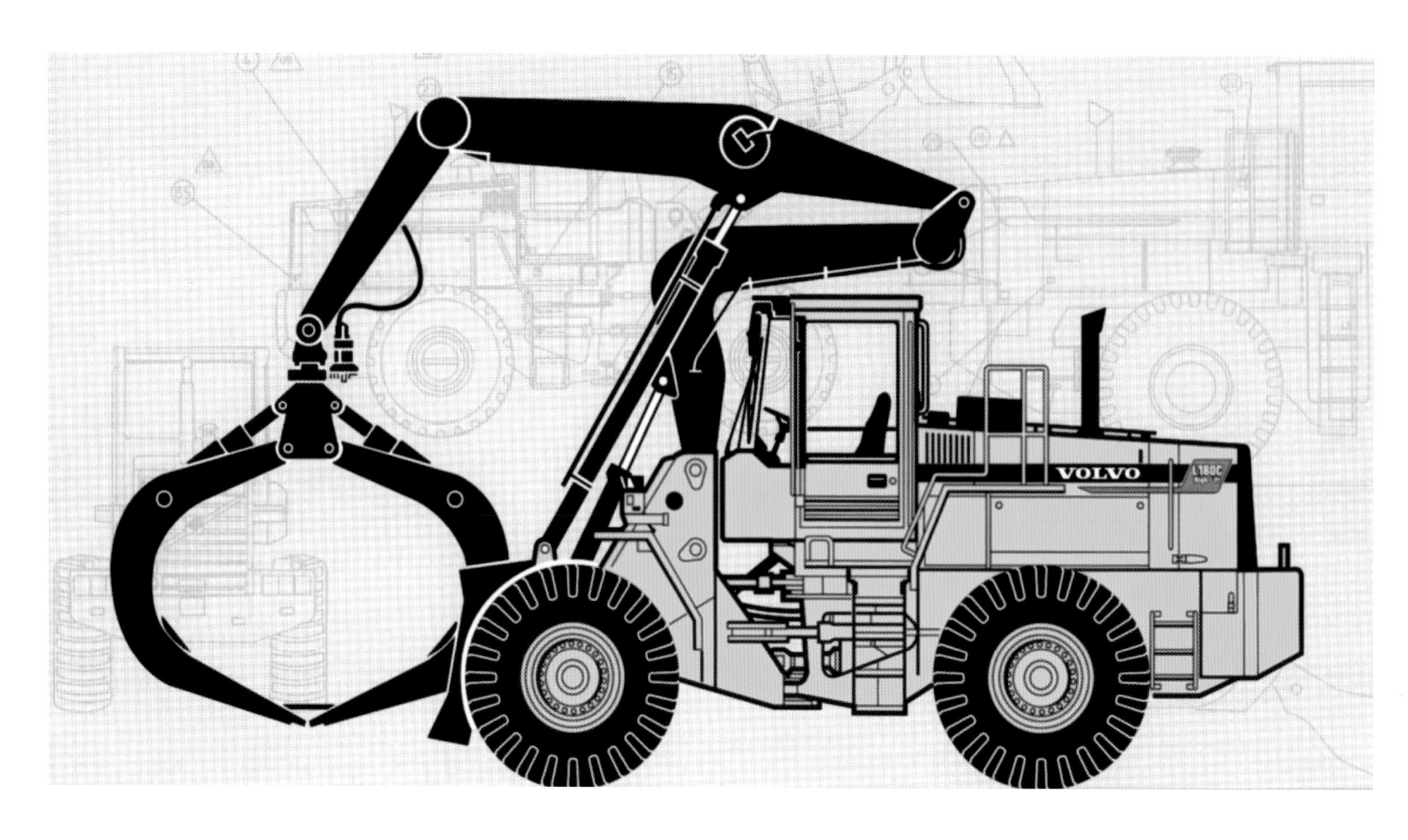


***The L180C Volvo wheeled loader in this configuration is intended for timber handling. The lift arm system is designed for high lift and long reach to enable effective operation and higher stacking. The grapple will swivel through 360 degrees.***

Two further variants of the wheeled loader theme for landfill sites are the Compactor and the Waste Handler. Volvo BM manufacture a compactor version of the L180C while Caterpillar produce landfill compactors based on their 816F, 826G, and 836 models. The major difference between these and standard wheeled loaders is obvious at a glance. The wheels fitted on the compactor variants are fitted not with pneumatic tires but with knife-shaped or trapeze pads designed to break up and compact the waste. Because of the potential destructive nature of the compactor's operating environment, there are a range of guards and protection plates available for the loader.

**VOLVO L180C WHEELED TIMBER HANDLER (HIGH-LIFT)**

**Engine**
Volvo TD 164 KHE

**Engine power (flywheel)**
209 kW (284 hp) @ 2100 rpm

**Bore and stroke**
n/a

**Transmission**
4F, 3R

**Maximum speed** (26.55R25 tires)

| | | |
|---|---|---|
| Forward | 21.81mph | 35.1 kph |
| Reverse | 14.91mph | 24 kph |

**Operating weight**
67,030lb 30,400 kg
(variable depending on grapple fitted)

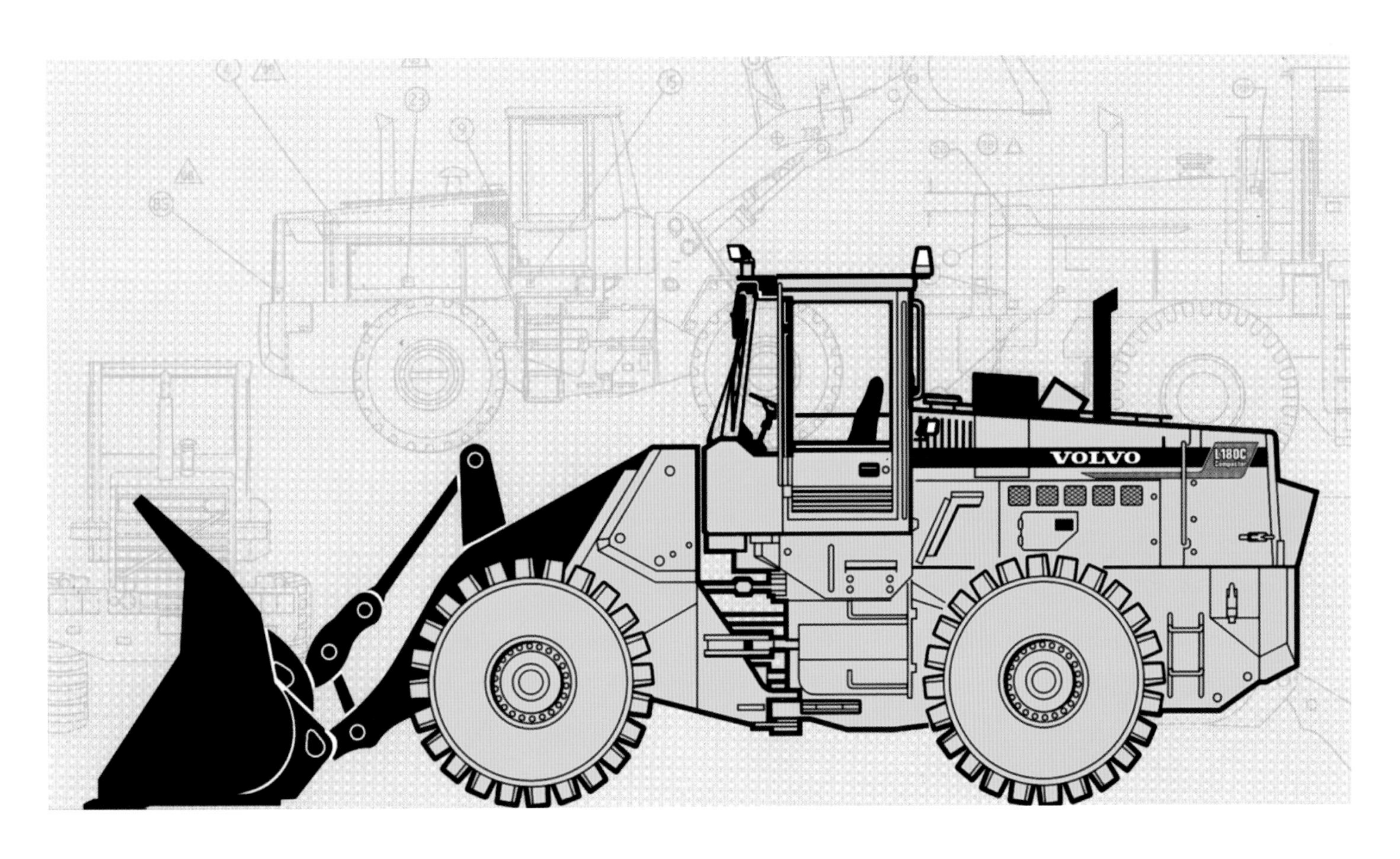


**VOLVO L180C WHEELED LOADER COMPACTOR**

**Engine**
Volvo TD 122 KHE

**Engine power (flywheel)**
209 kW (284 hp) @ 2100 rpm

**Bore and stroke**
n/a

**Transmission**
3F, 3R

**Maximum speed**

| | |
|---|---|
| Forward 5.16mph | 8.3 kph |
| Reverse 5.16mph | 8.3 kph |

**Bucket capacity**

| | |
|---|---|
| $6.02yd^3$ | $4.6\ m^3$ |

**Operating weight**

| | |
|---|---|
| 62,842.50lb | 28,500 kg |

***Because of its harsh operating environment, the L180C Volvo Compactor is available with a complete range of protection plates to prevent damage to the sides and underbody of the loader. These of course increase its operating weight.***

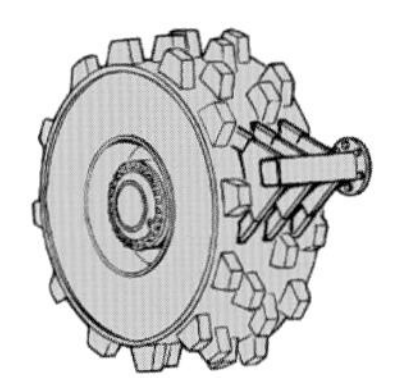
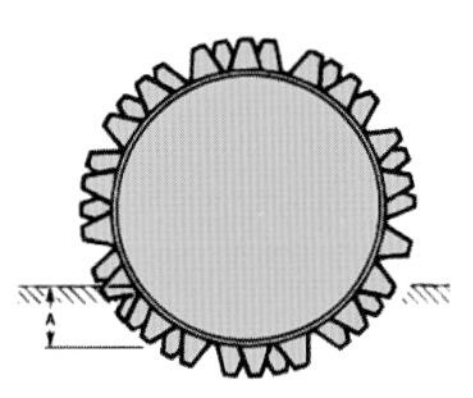

***A number of specific wheels have been designed for the Volvo L180C for use on landfill sites. These wheels feature a variety of pads in both knife and trapezoid shapes which are designed to enable the machine to operate on refuse where pneumatic tires would be punctured.***

VOLVO

| VOLVO BM L120C WASTE HANDLER | |
|---|---|
| **Engine** | |
| TD 73 KDE | |
| **Engine power (flywheel)** | |
| 153 kW (208 hp) @ 2100 rpm | |
| **Bore and stroke** | |
| n/a | |
| **Transmission** | |
| 4F, 3R | |
| **Maximum speed** (23.5R25 tires) | |
| Forward 22.06mph | 35.5 kph |
| Reverse 22.06mph | 35.2 kph |
| **Bucket capacity** | |
| 5.89yd$^3$ | 4.5 m$^3$ |
| **Operating weight** | |
| 50,759.10lb | 23,020 kg |

The Waste Handler looks more conventional in appearance but features a bucket designed specially for handling waste on landfill sites. It incorporates such items as a trash guard as well as windshield and underbody protection.

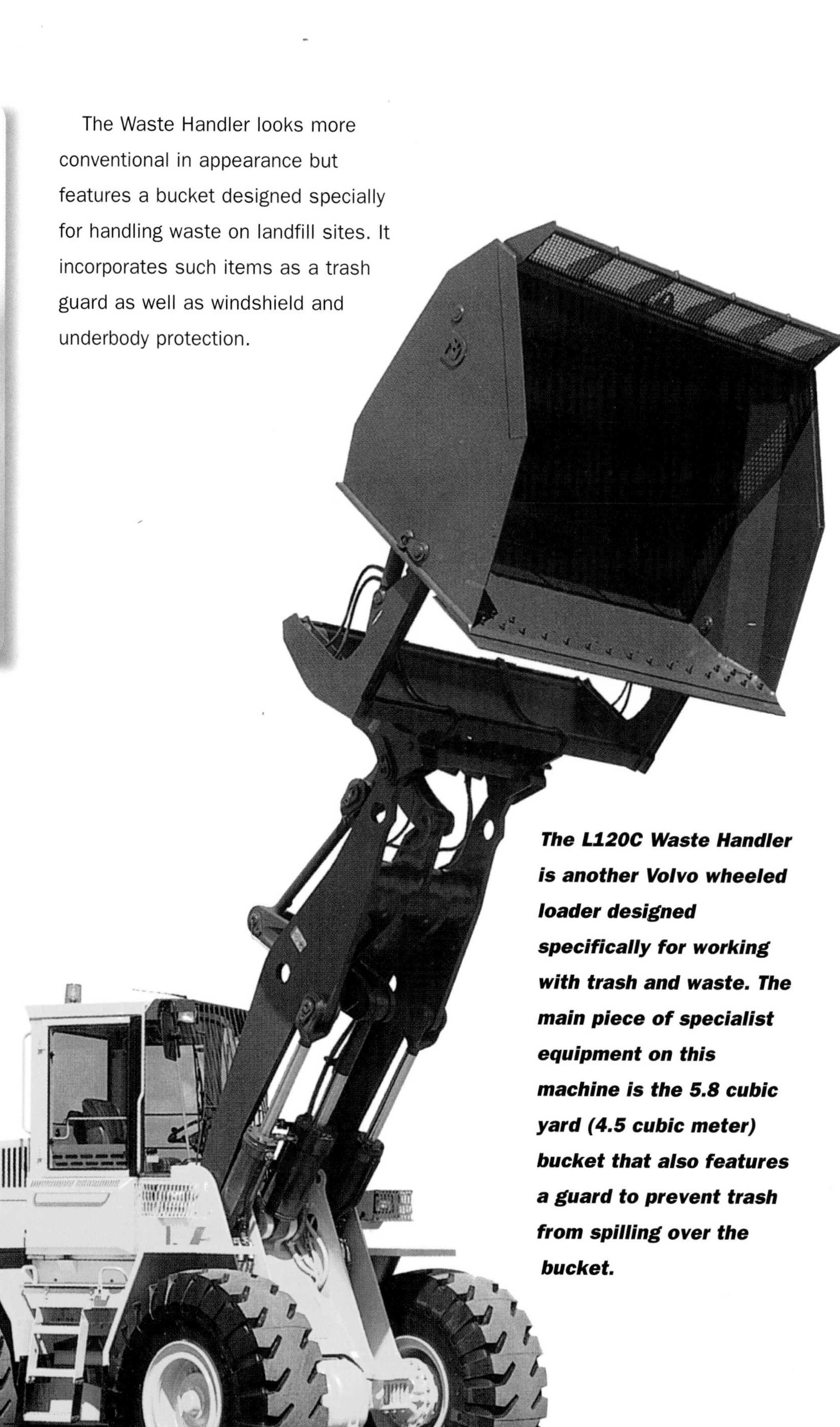


***The L120C Waste Handler is another Volvo wheeled loader designed specifically for working with trash and waste. The main piece of specialist equipment on this machine is the 5.8 cubic yard (4.5 cubic meter) bucket that also features a guard to prevent trash from spilling over the bucket.***

# Tracked Loaders

Tracked loaders are a further derivative of the wheeled loader but the machine is mounted on crawler tracks. Such loaders are considered suitable for a range of applications including heavy clearing, trench backfilling, rough dozing, land contouring, hard bank digging, and truck loading. Caterpillar manufacture a variety of such machines of which the 963B is typical. A box section main frame is the basis of the loader, comprising mild steel box section frame sections and castings used in areas of stress. To this are mounted the components of an oscillating undercarriage. The purpose of this is to keep both tracks in contact with the ground as much as possible (an improvement over the rigid

***The Caterpillar 963B Track Loader incorporates one of Cat's own engines, a hydrostatic drive system, an oscillating crawler undercarriage, and a Z-bar linkage around a box section main frame.***

**undercarriage design), reduce frame impact, and improve stability in rough terrain. The tracks are sealed and lubricated to reduce maintenance costs through bush and pin wear. This also reduces component friction for quieter running and greater drive efficiency. The sprocket rim segments are bolted on to allow replacement of worn segments without having to "break" the track. In the event that the track has to be dismantled–broken–a two-piece master link is included in the chain. The idler, which ensures track tension, is of a swing link design, allowing an amount of horizontal movement to maintain tension and eliminating the need for shims and wear strips.**

| CATERPILLAR 963B TRACK LOADER | |
|---|---|
| **Engine** | |
| Cat 3116 | |
| **Engine power (gross)** | |
| 130 kW @ 2000 rpm | |
| **Bore and stroke** | |
| 4.14 x 5.00in | 105 x 127 mm |
| **Transmission** | |
| Infinitely variable | |
| **Maximum speed** | |
| Forward 6.28mph | 10.1 kph |
| Reverse 6.28mph | 10.1 kph |
| **Bucket capacity** | |
| 3.01yd$^3$ | 2.3 m$^3$ |
| **Operating weight** | |
| 42,545.48lb | 19,295 kg |
| (variable depending on bucket fitted) | |
| **Ground pressure (17.73in/450 mm shoe)** | |
| 85.5 kPa | |
| **Chain links** | |
| 37 | |

The tracked loader is powered by a Cat 3116 diesel engine which is a six-cylinder, aftercooled, turbocharged, direct-injection design of engine. It is rear mounted which has the benefits of acting as a counterweight to the load in the bucket and allowing uninterrupted forward vision for the operator. The transmission is by means of hydrostatic drive, a hydraulic circuit for each track with drive pumps driven from the engine flywheel. The final drives are planetary isolated from the machine's weight and ground induced shock loads by the track roller frame pivots. Brakes are also hydrostatic.

The bucket is controlled by a Z-bar linkage which has been designed to be as simple as possible to minimize maintenance. Also minimizing this costly factor in loader operation is the use of sealed loader linkage points. The factors taken into account in the design of both the interior and exterior of the operator's cab are the same with wheeled loaders: air conditioning, ergonomic design, adjustable suspended seat, sound insulation, and rollover protection.

***The motive power is supplied by a Cat 3116 six cylinder, four cycle, direct injection, aftercooled, turbocharged diesel engine which has a high displacement to power ratio to ensure both longevity and reliability.***

Another major manufacturer who include a range of tracked loaders in

***The Liebherr LR641 crawler loader is the largest in that company's range.***

their range of earthmoving products is the German company Liebherr. The models range from the LR611 through the LR621C, LR622, LR631C, and LR641, the latter of which is the biggest machine.

The LR641 is powered by a direct-injection v-6, watercooled diesel engine which powers a closed-loop, hydrostatic drive transmission and steering system. These are oil cooled. Final drive is through three-stage spur gear drives. The sprockets feature replaceable segments and the chains double or triple grouser pads. The loader is a Z-bar linkage design hydraulically operated through a variable flow pump. Both travel and loader functions are controlled from within the cab by the operator through joystick levers. The cycle time is important to productivity and the bucket can be lifted in 7.8 seconds and lowered in 2.2 seconds; tilting out takes 1.8 seconds. To provide the ideal working environment, the cab is resiliently mounted, features a rollover protection system, and is equipped with pressurized air ventilation and an adjustable operator's seat to make operation as easy as possible.

**LIEBHERR LR641 CRAWLER LOADER**

**Engine**
Mercedes-Benz OM 441 A

**Engine power (gross)**
161 kW (219 hp) @ 2000 rpm

**Bore and stroke**
5.12 x 5.59in — 130 x 142 mm

**Transmission**
Infinitely variable

**Maximum speed**
Forward 6.21mph — 10 kph
Reverse 6.21mph — 10 kph

**Bucket capacity**
$3.79yd^3$ — $2.9\ m^3$

**Operating weight**
56,117.25lb — 25,450 kg
(variable depending on bucket fitted)

**Ground pressure (18.01in/457 mm shoe)**
13.51 PSI

**Chain links**
40

LIEBHE

# Rigid Dump Trucks

***The Aveling Barford RD40 is typical of the massive rigid dump trucks used in the mining industry with its scow end dump body and over cab protection.***

# Rigid Dump Trucks

**Rigid dump trucks are of a type that are also variously termed mining trucks and off-highway trucks. The idea behind the truck is straightforward: the rigid dump truck is a tipper-bodied, non-articulated, two-axle truck. What is less ordinary is the sheer size of them; they could easily be described as trucks on steroids. They are produced around the world by a list of companies that includes Aveling Barford, Caterpillar, Euclid, Heathfield Haulamatic, Komatsu, Liebherr, O&K, and Terex.**

The concept of off-highway dump trucks originated in the United States during the 1920s and 1930s. They came about because fleets of secondhand trucks that were pressed into service on many construction projects at that time soon wore out the tipping bodies, and the scow end tipper body began to be manufactured in steel. These bodies were transferred between trucks as the vehicles themselves wore out but by the mid-thirties completely new dump trucks were being fabricated by companies such as GMC, who produced a 6x4 model in 1934. A company known as the Euclid Road Machinery Company had been founded in 1931 as a division of the Euclid Crane and Hoist Company, and developed a series of off-highway dump trucks known as Trac-Trucks. Their first models were a 16.8-ton (15.24-tonne) rigid and a 22.40-ton (20.32-tonne) articulated.

In the postwar years companies around the globe produced off-highway rigid dump trucks. For example, in Finland Sisu produced the K-36 Mine Dumper in 1956. In England Foden had introduced their first specially designed dump truck in 1948. It was a tandem-drive 6x4 powered by the company's own FED6 two-stroke diesel engine with an eight-speed transmission. It had a steel scow body with a 10 cubic yard (7.65 cubic meter) capacity. Thorneycroft and Scammell, established British truck manufacturers, offered rigid dump truck models and by 1958 Foden offered a 56,000-pound (25,401.60-kilogram) payload dump truck. It was designated the FR.6/45 and was powered by a 300 brake horsepower, turbocharged Rolls Royce engine. This was later followed by the FC20.

Later companies such as Heathfield and Haulamatic developed rigid dump trucks including the 4-10 and the DF2D Transdumper from the respective manufacturers in the late 1960s. In Japan, Hino built the ZG13, a 175 brake horsepower, diesel-powered rigid dump truck, and in the United States the biggest payload machines were being built by International Harvester and WABCO. International Harvester supplied the 180 Payhauler which was driven by a two-stroke, 475 brake horsepower, V-12 Detroit Diesel and a ten-speed torque converter transmission. It had a capacity of 38 cubic yards (29.05 cubic meters). WABCO manufactured the Haulpak 150B rigid dump truck with a GM 16V-149T two-stroke, 16-cylinder diesel that produced 1,325 brake horsepower. The diesel electric truck had a heaped capacity of 97 cubic yards (74.17 cubic meters).

# Aveling Barford Rigid Dump Trucks

**Aveling Barford are a British company, a division of Wordsworth Holdings plc and based in Grantham, Lincolnshire. The company manufactures a range of five rigid dump trucks, the RD series. These are designated from RD30 to RD65, the numerical suffix reflects their payload in tons. Some of the major drivetrain components used in the Aveling Barford machines are proprietary items purchased from other major companies such as Cummins, Caterpillar, and Allison.**

***At 29.98 tons (27.2 tonnes) the RD30 has the smallest payload of Aveling Barford's British-made rigid dump trucks.***

The RD65 is the largest rigid dump truck in the range and is powered by a turbocharged and aftercooled in-line six-cylinder Cummins diesel. The transmission is an Allison ATEC6062 that has constant mesh gears in a planetary configuration and a TC683 torque converter. The gearshift is automatically electronically controlled and an air-operated hydraulic retarder, multiplate automatic lock-up clutch, and a downshift inhibitor are all fitted. The drive axle is manufactured by Aveling Barford and is of the double reduction type. The first reduction is at the crown wheel and pinion while the second is via planetary gearing in the wheel hubs. Fully floating half shafts connect the conventional differential assembly and the hubs. The complete axle is mounted to the chassis through a cast steel A-frame and the suspension units. The axle oscillates to allow wheels to stay in contact with the ground in uneven conditions. Braking is hydraulically actuated and the dual-circuit system uses open caliper disc brakes on all wheels.

The truck's frame is fabricated from high-tensile steel with an integral front bumper and stress-relieving castings at major support points on the machine. The heavy duty standard body has a wedge profile to ensure a clean discharge, it is of welded construction with strengthening members and underfloor ribs.

The steel cab is insulated for sound and temperature control. The operator's seat is a fully adjustable Bostrom item to increase operator comfort and the laminated windshield is angled forward to reduce glare.

| AVELING BARFORD RD30 RIGID DUMP TRUCK | |
|---|---|
| **Engine** | |
| Caterpillar 3406 DITA | |
| **Engine power** | |
| 261 kW (350 hp) @ 2100 rpm | |
| **Bore and stroke** | |
| 5.40 x 6.50in | 137 x 165 mm |
| **Transmission** | |
| 5F, 1R | |
| **Maximum speed** | |
| 31.19mph | 50.2 kph |
| **Tire size** | |
| 18.00-R25 | |
| **Wheelbase** | |
| 132.11in | 3,353 mm |
| **Length** | |
| 293.53in | 7,450 mm |
| **Maximum operating weight** | |
| 106,523.55lb | 48,310 kg |
| **Payload** | |
| 29.98t | 27.2 MT |

**Power for the 55 ton (45.5 tonne) capacity Aveling Barford RD55 comes from a six cylinder Cummins KTTA 19-C650 engine.**

**AVELING BARFORD RD55 RIGID DUMP TRUCK**

| | | |
|---|---|---|
| **Engine** | Cummins KTTA 19-C600 | |
| **Engine power** | 447 kW (600 hp) @ 2100 rpm | |
| **Bore and stroke** | 6.26 x 6.26in | 159 x 159 mm |
| **Transmission** | 6F, 1R | |
| **Maximum speed** | 32.31mph | 52 kph |
| **Tire size** | 21.00-R35 | |
| **Wheelbase** | 168.12in | 4,267 mm |
| **Length** | 361.69in | 9,180 mm |
| **Maximum operating weight** | 191,261.70lb | 86,740 kg |
| **Payload** | 50.04t | 45.4 MT |

**The RD65 is Aveling Barford's largest capacity dump truck. The capacity of the body heaped is 50 cubic yards (38.2 cubic meters) and truck capacity is 40 cubic yards (30.6 cubic meters).**

**AVELING BARFORD RD65 RIGID DUMP TRUCK**

| | | |
|---|---|---|
| **Engine** | Cummins KTTA 19-C700 | |
| **Engine power** | 522 kW (700 hp) @ 2100 rpm | |
| **Bore and stroke** | 6.26 x 6.26in | 159 x 159 mm |
| **Transmission** | 6F, 1R | |
| **Maximum speed** | 35.17mph | 56.6 kph |
| **Tire size** | 24.00-R35 | |
| **Wheelbase** | 169.10in | 4,292 mm |
| **Length** | 361.26in | 9,169 mm |
| **Maximum operating weight** | 228,217.50lb | 103,500 kg |
| **Payload** | 65.04t | 59 MT |

# Caterpillar Rigid Dump Trucks

**Peoria, Illinois based Caterpillar, produce a range of eight different trucks with payloads ranging from 39.68 tons (36 tonnes) to 240.30 tons (218 tonnes) in the 769D and 793B models respectively. Between these two come the 773D, 775D, and 777D, then the 785B and 789B machines both with payloads of over 110.23 tons (100 tonnes).**

***The Caterpillar 785B Mining Truck has a maximum operating weight of 550,000 pounds (249,480 kilograms) which means it operates in the 130 - 150 tons (118-136 tonne) class.***

**Caterpillar 785B**

At the heart of this monster truck is a Caterpillar 3512 EUI diesel engine of a four-stroke design. It has a long stroke to ensure sufficient torque, complete fuel combustion, and optimum efficiency. The engine has a large displacement and operates at low revolutions per minute to ensure long intervals between scheduled maintenance. The pistons within the engine are of a two-piece design intended to be durable and reduce exhaust emissions through improved combustion. The engine is electronically monitored and controlled. The transmission is a unit designed and manufactured by Caterpillar and the gearing is chosen to suit the high torque rise engine to minimize gear shifts in order to extend transmission life. The mechanical power train includes a lock-up torque converter that engages at low speeds in first gear to provide high rimpull, and a lock-up clutch that engages at approximately five miles per hour (8.05 kilometers per hour) to operate more efficiently in direct drive. The transmission is six speed and features a planetary power shift and incorporates large-diameter clutches which are individually modulated for fast, smooth gear shifts. The transmission oil is contained in a separate reservoir than other oils to prevent contamination. The transmission is electronically controlled through use of EPTC II–Electronic Programmable Transmission Control–which uses electronically transferred engine revolutions per minute data to change gear at factory preset points to optimize performance. Top gear, downshift, and anti-hunt control functions are also electronically

***The 785B is powered by the 3512 EUI diesel engine which is a unit of four stroke design with a long stroke to maximize fuel combustion.***

| CATERPILLAR 785B MINING TRUCK | |
|---|---|
| **Engine** | |
| Cat 3512 EUI | |
| **Engine power** | |
| 1029 kW (1,380 hp) @ 1750 rpm | |
| **Bore and stroke** | |
| 6.70 x 7.49in | 170 x 190 mm |
| **Transmission** | |
| 6F, 1R | |
| **Maximum speed** | |
| 34.05mph | 54.8 kph |
| **Tire size** | |
| 33.00-R51 (E4) | |
| **Wheelbase** | |
| 204.09in | 5,180 mm |
| **Length** | |
| 434.27in | 11,022 mm |
| **Maximum operating weight** | |
| 550,103.40lb | 249,480 kg |
| **Payload** | |
| 130.07t | 118 MT |

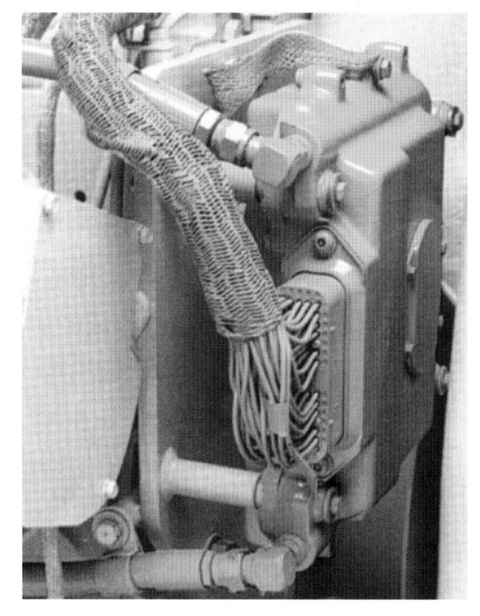

***The engine functions are electronically controlled, a factor that minimizes fuel consumption through monitoring operator and sensor inputs.***

controlled. The latter function is to stop the gearbox hunting for gears on a gradient. The electronic controls of the transmission allow for subsequent diagnosis of faults and problems by service personnel.

The integration of the engine and transmission is comprehensive: engine revolutions per minute are regulated during gearshifts, engine speed is regulated during forward/reverse directional shifts, there is a neutral coast inhibitor which prevents the transmission from shifting to neutral at speeds in excess of four miles per hour (6.44 kilometers per hour) to protect the transmission, and there is a body-up shift inhibitor to stop the transmission shifting above a preprogrammed gear without the body being fully lowered. Another electronically controlled function is the automatic electronic traction aid where if slippage of a wheel exceeds a certain limit in wet or difficult conditions the oil-cooled disc brakes engage and slow the spinning wheel. Torque is then automatically

| CATERPILLAR 773D OFF-HIGHWAY TRUCK | |
|---|---|
| **Engine** | |
| Cat 3412E | |
| **Engine power** | |
| 509 kW (682 hp) @ 2000 rpm | |
| **Bore and stroke** | |
| 5.40 x 5.99in | 137 x 152 mm |
| **Transmission** | |
| 7F, 1R | |
| **Maximum speed** | |
| 25.54mph | 41.1 kph |
| **Tire size** | |
| 24.00-R35 (E4) | |
| **Wheelbase** | |
| 165.13in | 4,191 mm |
| **Length** | |
| 362.87in | 9,210 mm |
| **Maximum operating weight** | |
| 204,028.65lb | 92,530 kg |
| **Payload** | |
| 57.65t | 52.3 MT |

***The Caterpillar 773D off-highway truck is designed for low cost hauling in mining and construction.***

***The Caterpillar 777C rigid dump truck.***

transferred to the other wheels with better traction.

Braking is done by oil-cooled, multiple-disc brakes at each wheel and engine retarding where the engine runs against compression on downhill hauls. Cooling of the discs is continuous to eliminate brake fade. Retarding is split 65/35 percent rear/front to provide control in slippery conditions. The Automatic Retarder Control (ARC) electronically controls braking on grade to maintain the engine speed at approximately 1900 revolutions per minute. While the ARC modulates the brakes, the operator may apply additional braking force using either the manual retarder or the brake pedal. When the operator applies the throttle the ARC is overridden. The ARC has a facility to activate automatically when engine speeds exceed preset factory limits to prevent damage. The benefits of ARC are increased production with faster downhill speeds, excellent control and reduced operator effort, extended component life from reduced torque fluctuations in the brake system, and faster trouble shooting and diagnosis.

The backbone of the whole truck is its being based around a box-section design structure manufactured in mild steel and a number of other components including two forgings and 21 castings. The castings with large

| CATERPILLAR 785D QUARRY TRUCK | |
|---|---|
| **Engine** | |
| Cat 3412E | |
| **Engine power** | |
| 541 kW (725 hp) @ 2000 rpm | |
| **Bore and stroke** | |
| 5.40 x 5.99in | 137 x 152 mm |
| **Transmission** | |
| 7F, 1R | |
| **Maximum speed** | |
| 40.89mph | 65.8 kph |
| **Tire size** | |
| 24.00-R35 (E4) | |
| **Wheelbase** | |
| 165.13in | 4,191 mm |
| **Length** | |
| 366.42in | 9,300 mm |
| **Maximum operating weight** | |
| 235,039.77lb | 106,594 kg |
| **Payload** | |
| 65.37t | 59.3 MT |

***Caterpillar mining trucks such as the 789B are complex machines completely integrated in their design using a mechanical power train that is electronically controlled.***

| CATERPILLAR 789B MINING TRUCK | |
|---|---|
| **Engine** | |
| Cat 3516 | |
| **Engine power** | |
| 1342 kW (1,800 hp) @ 1750 rpm | |
| **Bore and stroke** | |
| 6.70 x 7.49in | 170 x 190 mm |
| **Transmission** | |
| 6F, 1R | |
| **Maximum speed** | |
| 33.80mph | 54.4 kph |
| **Tire size** | |
| 37.00-R57 (E4) | |
| **Wheelbase** | |
| 224.58in | 5,700 mm |
| **Length** | |
| 479.77in | 12,177 mm |
| **Maximum operating weight** | |
| 700,131.60lb | 317,520 kg |
| **Payload** | |
| 169.76t | 154 MT |

radii are used in areas of high stress to eliminate fatigue and cracking. The advantage of using mild steel for the box section frame is that it provides flexibility, durability, and resistance to impact loads even in cold climates. The Rollover Protection Structure (ROPS) is integral to the cab and frame. The truck body is both rugged and durable to provide service in the toughest of mining applications. It features a dual-slope floor designed to provide good load retention, low center of gravity, and clean dumping characteristics. To strengthen the body, box-beams are integral to the floor, sidewalls, top rail, corner, and cab canopy areas, providing added impact resistance. The top rail down each side of the body is reinforced rolled steel. This both increases body strength and protects the machine from damage from falling material from the loading tool. The V-section bottom reduces the shock loading and centers the load, while the forward body slope and rear ducktail slope help to retain loads when climbing steep gradients.

The operating station, as the cab is known, is ergonomically designed to create a safe, productive, and comfortable environment. The ROPS sound-suppressed cab is resiliently mounted to minimize vibration, and visibility is intended to be excellent all-round. The operator's seat is fully adjustable and air-suspended. The various management systems have displays in front of the operator and the controls are designed to come easily to hand.

# Euclid Rigid Haulers

**Euclid invented the off-highway hauler in 1926 and since then the company's haulers have earned a reputation as some of the strongest and most durable machines in the industry. Euclid's products are engineered for heavy work around the clock with a minimum of downtime for maintenance. The machines are based on Euclid's tough frame to which a variety of power trains and bodies are added, along with a sophisticated suspension system.**

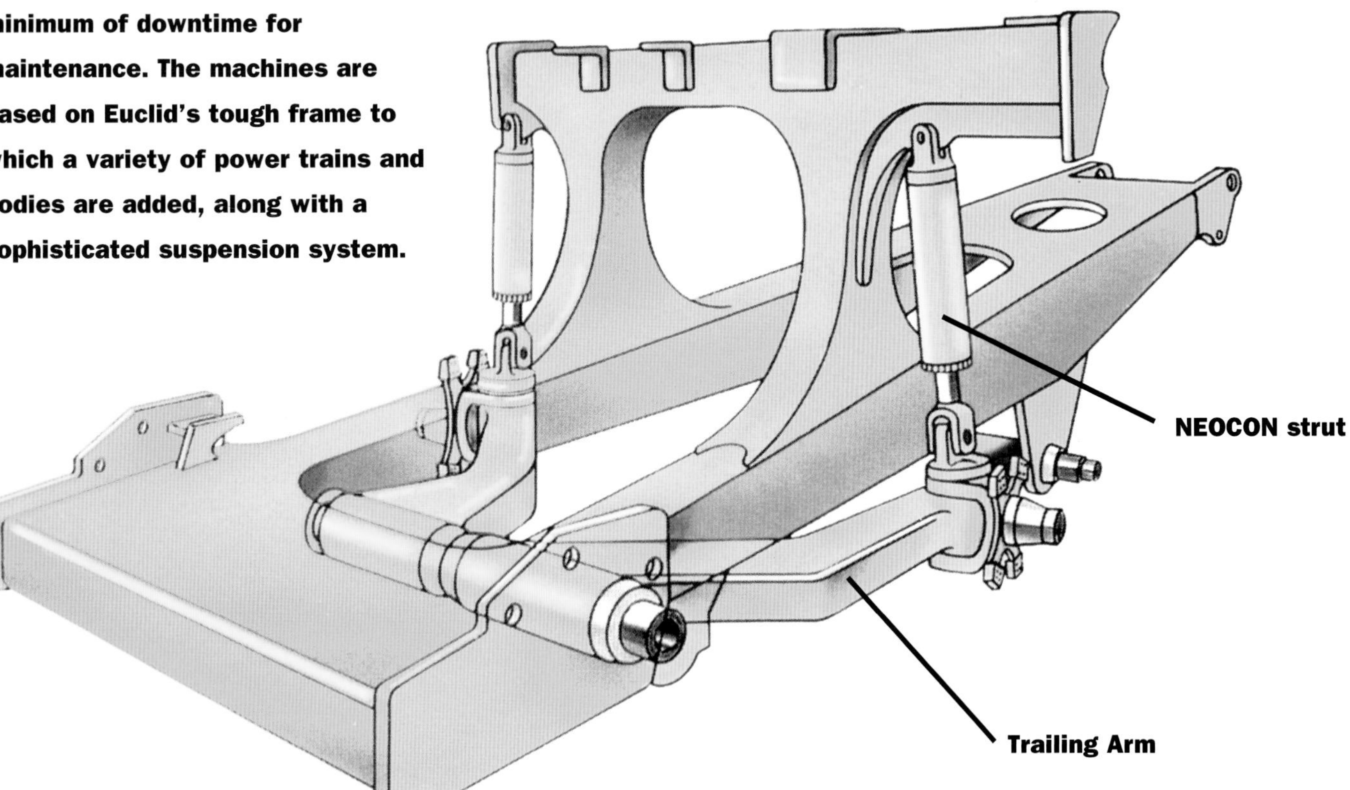


***Euclid's rigid dump trucks incorporate features such as ACCU-TRAC suspension which features trailing arms and NEOCON struts for the front wheels.***

Currently the company, now known as Euclid-Hitachi Heavy Equipment, offers a complete range of rigid dump trucks with nominal capacities ranging from 27.56 to 209.44 tons (25 to 190 tonnes). The scale of work undertaken by trucks such is these is illustrated by the following. In 1970, 54 Euclid 117.60-ton (106.69-tonne) haulers were used to open the Bougainville Copper Limited mine in Papua New Guinea. The machines worked on the same mine until 1977. In 1985 another fleet of Euclid haulers, 45 190.40-ton (172.73-tonne) machines, had worked in excess of 3 million hours.

***Euclid's own design of wet disc brake is engineered for long service life even in extreme environments. Wet discs are fitted to the rear axle and are intended for use for service and secondary braking as well as for retarding.***

| EUCLID R60 RIGID DUMP TRUCK | |
|---|---|
| **Engine** | |
| Cummins QSK19-C700 | |
| **Engine power** | |
| 522 kW (700 hp) @ 2100 rpm | |
| **Bore and stroke** | |
| 6.26 x 6.26in | 159 x 159 mm |
| **Transmission** | |
| 6F, 2R | |
| **Maximum speed** | |
| 35.98mph | 57.9 kph |
| **Tire size** | |
| 24.00-R35 (E3) | |
| **Wheelbase** | |
| 169.03in | 4,290 mm |
| **Length** | |
| 366.42in | 9,300 mm |
| **Maximum operating weight** | |
| 211,680lb | 96,000 kg |
| **Payload** | |
| 63.05t | 57.2 MT |

***The R60 machine is powered by a Cummins engine.***

| EUCLID R65 RIGID DUMP TRUCK | |
|---|---|
| **Engine** | |
| Cummins VTA28-C | |
| **Engine power** | |
| 567 kW (760 hp) @ 2100 rpm | |
| **Bore and stroke** | |
| 5.52 x 5.99in | 140 x 152 mm |
| **Transmission** | |
| 6F, 2R | |
| **Maximum speed** | |
| 22.00mph | 35.4 kph |
| **Tire size** | |
| 24.00-R35 (E3) | |
| **Wheelbase** | |
| 169.03in | 4,290 mm |
| **Length** | |
| 366.42in | 9,300 mm |
| **Maximum operating weight** | |
| 225,037.89lb | 102,058 kg |
| **Payload** | |
| 67.68t | 61.4 MT |

***The R65 features hydraulically actuated front and rear brakes.***

**R60 and R65 Dump trucks**

Euclid's smaller-capacity dump trucks such as the R60 and R65 models, feature the Command cab which has integral ROPS and, through its double-wall construction, controls both sound and temperature within it. The whole unit is rubber iso-mounted to the chassis through a three-point fixing to minimize vibration. Mechanically the machines are advanced, featuring hydraulically actuated front dry disc brakes and wet disc rear brakes. Cummins engines and Allison

***The R190 has options of engine and tire type depending upon the task to be undertaken.***

transmissions are proprietary units and drive through a Euclid Model 2350 rear axle which features reductions in both the differential and in the planetary gears in the hubs.

**Euclid R190 Dump truck**

This massive machine features a payload in the range of 189.93 to 209.44 tons (172.3 to 190 tonnes) and has options of various items including engine and tire types. Either Cummins or Detroit Diesel engines are supplied and different tire sizes are available depending on usage. The transmission is different than the other Euclid products because it is electric rather than mechanical drive. This system uses controls, alternator, and Wheel Motors from General Electric–the Statex SSL, GTA-22F, and 788BS units respectively. These components are cooled by a radiator, fan, and blower which are mounted on a subframe within the main frame.

**EUCLID**

| EUCLID R190 RIGID DUMP TRUCK | |
|---|---|
| **Engine** | |
| Cummins KTTA50-C | |
| **Engine power** | |
| 1342 kW (1,800 hp) @ 1900 rpm | |
| **Bore and stroke** | |
| 6.26 x 6.26in | 159 x 159 mm |
| **Transmission** | |
| Electric drive | |
| **Maximum speed** | |
| 32.69mph | 52.6 kph |
| **Tire size** | |
| 36.00-R51 RL 4H | |
| **Wheelbase** | |
| 222.22in | 5,640 mm |
| **Length** | |
| 464.53in | 11,790 mm |
| **Maximum operating weight** | |
| 1,428,840lb | 648,000 kg |
| **Payload** | |
| 189.60t | 172 MT |

# Heathfield Haulamatic Ltd. Rigid Dump Trucks

**HEATHFIELD H44 RIGID DUMP TRUCK**

| | | |
|---|---|---|
| **Engine** | Cummins KTA19-C | |
| **Engine power** | 358 kW (480 hp) @ 2100 rpm | |
| **Bore and stroke** | 6.26 x 6.26in | 159 x 159 mm |
| **Transmission** | 6F, 1R | |
| **Maximum speed** | 41.01mph | 66 kph |
| **Tire size** | 18.00-R33 | |
| **Wheelbase** | 141.84in | 3,600 mm |
| **Length** | 308.70in | 7,835 mm |
| **Maximum operating weight** | 162,552.60lb | 73,720 kg |
| **Payload** | 44.09t | 40 MT |

**HEATHFIELD H50 RIGID DUMP TRUCK**

| | | |
|---|---|---|
| **Engine** | Cummins KTA19-C | |
| **Engine power** | 391 kW (525 hp) @ 2100 rpm | |
| **Bore and stroke** | 6.26 x 6.26in | 159 x 159 mm |
| **Transmission** | 6F,1R | |
| **Maximum speed** | 40.01mph | 66 kph |
| **Tire size** | 21.00-R35 | |
| **Wheelbase** | 141.84in | 3,600 mm |
| **Length** | 308.70in | 7,835 mm |
| **Maximum operating weight** | 177,105.60lb | 80,320 kg |
| **Payload** | 49.60t | 45 MT |

***The Heathfield H44 is powered by a Cummins diesel engine and uses an Allison transmission.***

***The Heathfield H50 seen here in a typical working environment was launched in the spring of 1997 and has a larger payload than the H44S.***

**Heathfield Haulamatic Ltd. are based in England and are a subsidiary of L.H.Group Holdings.**

They manufacture rigid dump trucks such as the H44 and H50 models. Both models use proprietary engines including Cummins engines and Allison transmissions, namely the KTA19-C and CLBT 5963 ATEC units respectively. The engine is of the turbocharged, aftercooled, in-line, six-cylinder type while the transmission is six speed.

# Liebherr Mining Trucks

**Liebherr Mining Truck Inc. is a subsidiary of Liebherr-America Inc. and acquired the mining truck business and production capability of Wiseda in mid-1995. Wiseda manufactured massive rigid dump trucks such as "The King of the Lode" and introduced the first 268.80-plus ton (243.86-plus tonne) capacity, two-axle, diesel-electric, rear-dump haulage truck in 1982. The company gained a reputation for building innovatively designed trucks with low operating costs. The buy out of Wiseda by Liebherr surprised some in the industry because Liebherr traditionally develops new products in-house. It was seen by others as Liebherr moving with the changing times in the equipment market where many companies prefer to buy all their machines from one manufacturer for a specific project. To be competitive in this way Liebherr, having acquired Wiseda, could offer a more comprehensive product range. Wiseda were a Kansas-based company who manufactured only one product, the 240.30-ton (218-tonne) capacity KL-2450 Mining Truck. Production per annum was in single figures–nine KL2450 models in the year of the buy out–and with Liebherr production was to be almost 30 trucks per year. The company planned to produce 50 per year subject to sufficient market demand and brought some of its European specialists to Baxter Springs in Kansas to increase production capability.**

***The Liebherr KL-2450 Mining Truck is of the diesel electric type.***

A new model from the company was the KL-2420, a scaled-down version of the KL-2450, and work is ongoing on a 352.74-ton (320-tonne) capacity truck–the KL-2640–to compete with Komatsu-Haulpak's 930 model. This hauler has many of the design principles of the previous Liebherr haulers but features a new AC drive system and incorporates dynamic retarding. Like Liebherr's other dump trucks the KL-2640 is a two-axle, six-tire, rear–dump hauler with a box-section main frame and multiple cross members. The engine package rolls out and the deck is removable as a unit with the cab, control box, and grid box mounted as they are on other Liebherr haulers.

| LIEBHERR KL-2450 MINING TRUCK | |
|---|---|
| **Engine** | |
| Detroit diesel 16V149 | |
| **Engine power** | |
| 1343 kW (1,800 hp) | |
| **Bore and stroke** | |
| n/a | |
| **Transmission** | |
| Electric drive | |
| **Maximum speed** | |
| n/a | |
| **Tire size** | |
| 37-R57 (E4) | |
| **Wheelbase** | |
| 228.52in | 5,800 mm |
| **Length** | |
| 524.02in | 13,300 mm |
| **Maximum operating weight** | |
| 28,444.50lb | 12,900 kg |
| **Payload** | |
| 189.60t | 172 MT |

***A KL-2450 removing overburden at the Martiki Mine in Lovely, Kentucky, USA.***

# Trucks at Work

***The Martiki Mine uses a fleet of five KL-2450 haulers in its operations.***

**KL-2450 Trucks at work**

Mining operations where Liebherr KL-2450 Mining Trucks are in operation give an indication of the scale of mining being undertaken and underline the need for 240.30-ton (218-tonne) haulers. In the central Appalachian area of the United States the Martiki Mine in Lovely, Kentucky, is removing a mountain top to extract coal through combined use of dragline, truck, and shovel. Shovels open up an area by contour mining around peaks and recovering the outcrop of the coal seams which vary between 48 and 72 inches (121.92 and 182.88 centimeters) thick. The Martiki Mine

***Peabody Coal mine in Grants, New Mexico, USA and use five KL-2450 haulers for removing both coal and overburden.***

produces over 2.8 million tons (2.54 million tonnes) of coal per annum and ships it in six trains per week. For overburden material removal the mine management uses a fleet of five Liebherr KL-2450 trucks fitted with Detroit diesels and the General Electric 787 Statex drive system.

In the Powder River Basin near Wright, Wyoming, is situated Kerr McGee's Jacobs Ranch Mine which is strip mining coal. It is the third-largest coal producer in the United States, and produced 27,602,522 tons (25,041,008 tonnes) in 1995. Coal is mined for power-generating purposes and shipped by train daily. A fleet of KL-2450 haulers, and other equipment, is used for both over-burden and coal hauling on the 24-hour schedule operated at the Jacobs Ranch Mine. Winter in the area is hard and the trucks are regularly expected to operate at temperatures as low as -30 degrees Fahrenheit (-34.4 degrees centigrade) and in feet of accumulated snow. Another mine in the same general area is the Black Thunder Mine in the vicinity of Gilette, Wyoming, where low-sulfur coal is extracted. The primary customers for the coal are electric utilities, and the Black Thunder Mine has been the United States' top producer of coal since 1982. It started as a truck and shovel operation and has increased dramatically in size since opening. Thirteen KL-2450 diesel electric haulers are amongst the machinery employed on this undertaking, the first of which was purchased in 1986.

Peabody Coal operate the Lee Ranch Coal Company in Cibola County, near Grants New Mexico, where coal for power generation has been extracted since 1984. It is a strip mine using draglines and truck/shovel combinations, and coal is shipped daily by train. The coal seams range from 24 to 60 inches (50.8 to 152.4 centimeters) thick, and both coal and overburden are moved in a fleet of five KL-2450 haulers loaded by a 56 cubic yard (42.82 cubic meter) bucket shovel.

It is not only in coal mining that Liebherr haulers are used; in North Central Nevada is the Echo Bay Mines McCoy/Cove Mine where gold is extracted. The mine produces in excess of 330,000 ounces (9,355,500 grams) of gold and 8,000,000 ounces (22,680,000 grams) of silver each year. To extract these minerals, massive amounts of both ore and overburden must be moved and the mine uses a fleet of four diesel electric KL-2450 haulers and various other mechanical trucks.

# Orenstein & Koppel AG

***The O&K K45 is powered by a Cummins diesel engine and its designation refers to its payload in tons.***

**O&K make a range of five rigid dump trucks of varying payloads that range between 68,355 and 200,037.60lb (31,000 and 90,720 kg). The company is one of Germany's oldest-established engineering firms and was founded in Berlin by Benno Orenstein and Arthur Koppel in 1876. Among the company's earliest products were steam excavators but in the postwar years the emphasis has been on construction and strip-mining machines, such as the K-Series of rigid dump trucks. These trucks, like some from other manufacturers, use a variety of proprietary components including Cummins engines and Allison transmissions.**

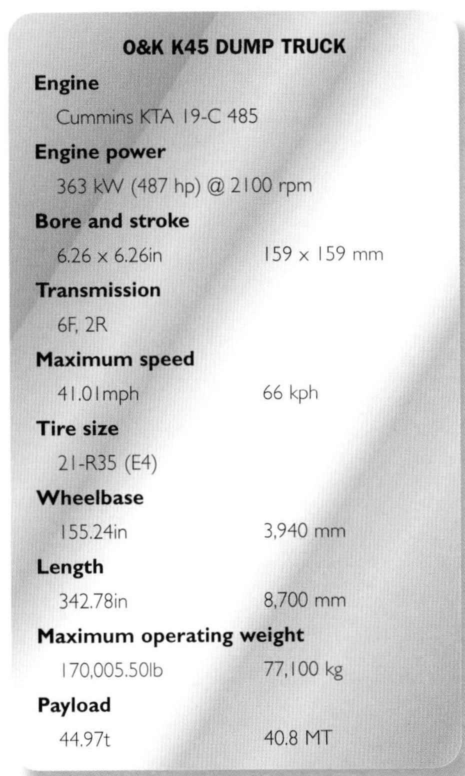

**O&K K45 DUMP TRUCK**

| | | |
|---|---|---|
| **Engine** | | |
| | Cummins KTA 19-C 485 | |
| **Engine power** | | |
| | 363 kW (487 hp) @ 2100 rpm | |
| **Bore and stroke** | | |
| | 6.26 x 6.26in | 159 x 159 mm |
| **Transmission** | | |
| | 6F, 2R | |
| **Maximum speed** | | |
| | 41.01mph | 66 kph |
| **Tire size** | | |
| | 21-R35 (E4) | |
| **Wheelbase** | | |
| | 155.24in | 3,940 mm |
| **Length** | | |
| | 342.78in | 8,700 mm |
| **Maximum operating weight** | | |
| | 170,005.50lb | 77,100 kg |
| **Payload** | | |
| | 44.97t | 40.8 MT |

***The O&K K60 rigid dump truck is driven by a 504 kW Cummins diesel engine. It has a payload of 59.97 tons (54.5 tonnes).***

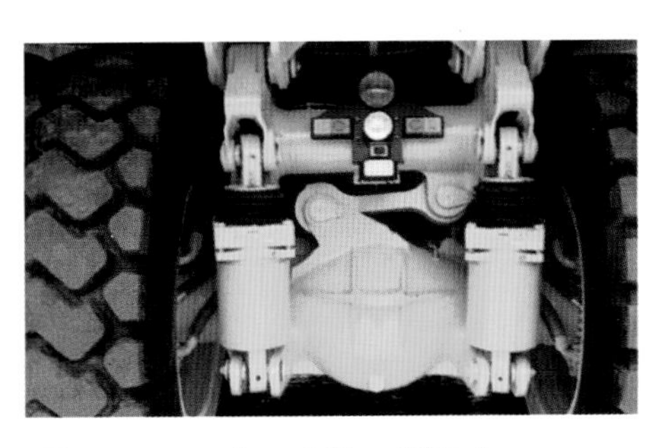

***The rear axle of the K60 is attached to the frame by means of an A-frame and a transverse link.***

***Because of the abrasive materials to be carried, O&K Hardox 400 steel is used in the construction of the dump body.***

**O&K K60 DUMP TRUCK**

| | | |
|---|---|---|
| **Engine** | | |
| | Cummins KTA 19-C 675 | |
| **Engine power** | | |
| | 504 kW (675 hp) @ 2100 rpm | |
| **Bore and stroke** | | |
| | 6.26 x 6.26in | 159 x 159 mm |
| **Transmission** | | |
| | 6F, 2R | |
| **Maximum speed** | | |
| | 35.42mph | 57 kph |
| **Tire size** | | |
| | 24-R35 XHD1A (E4) | |
| **Wheelbase** | | |
| | 155.24in | 3,940 mm |
| **Length** | | |
| | 351.45in | 8,920 mm |
| **Maximum operating weight** | | |
| | 210,357lb | 95,400 kg |
| **Payload** | | |
| | 59.97t | 54.4 MT |

***The K100 being loaded; for maximum performance the capacity of the truck and loader must be matched in order to minimize waiting times.***

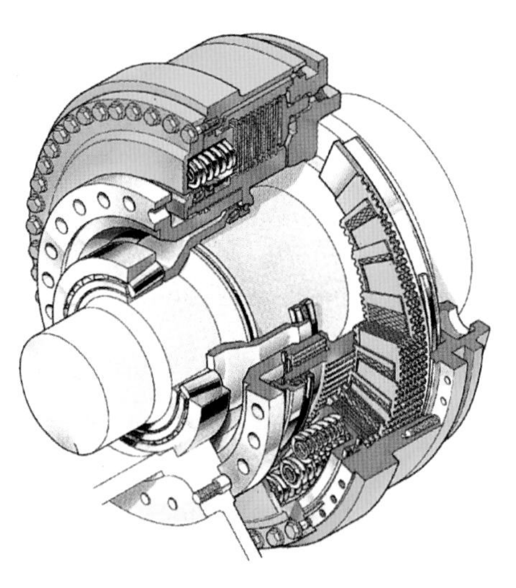

***A section through the oil cooled multi-plate rear hub braking assembly of the O&K rigid dump truck series.***

O&K's rigid dump trucks are completely at home in the deepest quarries and open cast mines but are also frequently deployed on construction projects where large quantities of soil and rock require removal. O&K dump trucks have, for example, recently been used for highway construction in both Greece and Spain as well as in construction of the new airport in Hong Kong.

The dump truck range includes models with artic steering as well as rigid frame types. The artic-steered type have all-wheel drive and make use of this capability in the difficult terrain of mines. The rigid frame models have higher payloads and are designed for maximum handling performace.

| O&K K100 DUMP TRUCK | |
|---|---|
| **Engine** | |
| Cummins KTA 38-C 1050 | |
| **Engine power** | |
| 783 kW (1,050 hp) @ 2100 rpm | |
| **Bore and stroke** | |
| 6.26 x 6.26in | 159 x 159 mm |
| **Transmission** | |
| 6F, 2R | |
| **Maximum speed** | |
| 40.39mph | 62 kph |
| **Tire size** | |
| 27-R49 XHD1A | |
| **Wheelbase** | |
| 180.06in | 4,570 mm |
| **Length** | |
| 426.31in | 10,820 mm |
| **Maximum operating weight** | |
| 340,011lb | 154,200 kg |
| **Payload** | |
| 106.92t | 97 MT |

*The Terex 3345 Off-Highway truck has a maximum payload of 45 tons (41 tonnes) and a heaped capacity of 34 cubic yards (26 cubic meters).*

# Terex Off-Highway Trucks

**Terex have plants in Motherwell, Scotland, and Tulsa, Oklahoma, where they construct a range of heavy equipment including rigid dump trucks. There are six trucks in the range, of varying payloads, but all are variations on the rigid-framed, rear-dumping tipper truck. The exact components used vary from truck to truck, depending on the requirements. The largest in the range, the 33100 Mining Truck, is based on a box-section frame that includes an integral front bumper, and relies on the added strength of castings in high-stress areas. The front suspension struts are bolted to a closed loop member of the frame and the high-capacity body is hinged to the rear of the frame. The body has a 74.56 cubic yard (57 cubic meter) heaped volume capacity and is made from abrasion-resistant steel. It has a uniform depth from front to rear to match the capacities of wheeled loaders but features an angled tail chute to retain the load and to ensure controlled dumping. The body is heated by the exhaust to ensure shedding of the load.**

The truck features a Cummins diesel engine and an Allison ATEC (automatic electronic control) transmission. Drive is mechanical, with a dual reduction through the differential and planetary gears. Brake discs are oil cooled at the rear and dry at the front.

| TEREX 3345 OFF HIGHWAY TRUCK | |
|---|---|
| **Engine** | |
| Cummins KTA 19-C | |
| **Engine power** | |
| 363 kW (487 hp) @ 2100 rpm | |
| **Bore and stroke** | |
| 6.26 x 6.26in | 159 x 159 mm |
| **Transmission** | |
| 6F, 2R | |
| **Maximum speed** | |
| 40.95mph | 65.9 kph |
| **Tire size** | |
| 21-R35 radial | |
| **Wheelbase** | |
| 155.24in | 3,940 mm |
| **Length** | |
| 342.78in | 8,700 mm |
| **Maximum operating weight** | |
| 170,027.55lb | 77,110 kg |
| **Payload** | |
| 45.19t | 41 MT |

***A Terex 3360 with a load approaching the truck's heaped capacity of 46 cubic yards (35 cubic meters).***

***A TEREX frame.***

***The TEREX multi-plate rear axle brake.***

| TEREX 3360 OFF HIGHWAY TRUCK | |
|---|---|
| **Engine** | |
| Cummins KTTA 19-C | |
| **Engine power** | |
| 504 kW (675 hp) @ 2100 rpm | |
| **Bore and stroke** | |
| 6.26 x 6.26in | 159 x 159 mm |
| **Transmission** | |
| 6F, 2R | |
| **Maximum speed** | |
| 35.42mph | 57 kph |
| **Tire size** | |
| 24-R35 radial | |
| **Wheelbase** | |
| 155.24in | 3,940 mm |
| **Length** | |
| 351.45in | 8,920 mm |
| **Maximum operating weight** | |
| 210,423.15lb | 95,430 kg |
| **Payload** | |
| 60.63t | 55 MT |

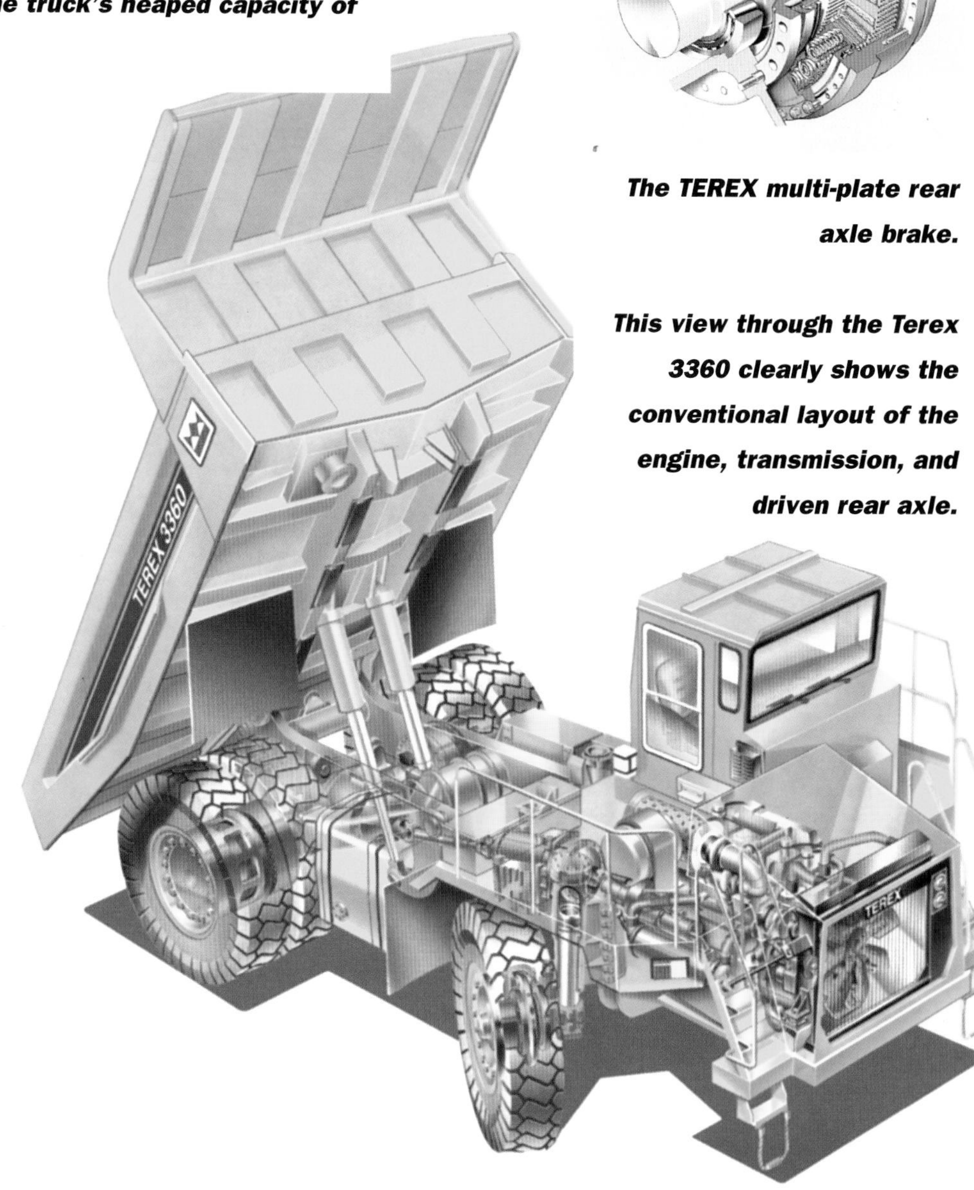


***This view through the Terex 3360 clearly shows the conventional layout of the engine, transmission, and driven rear axle.***

The 100 ton (91 tonne) payload Terex 33100 Mining Truck.

*The well appointed operator station within the Terex 33100 Mining Truck.*

**TEREX 33100 MINING TRUCK**

| | | |
|---|---|---|
| **Engine** | Cummins KTA 38-C | |
| **Engine power** | 783 kW (1,050 hp) @ 2100 rpm | |
| **Bore and stroke** | 6.26 x 6.26in | 159 x 159 mm |
| **Transmission** | 6F, 1R | |
| **Maximum speed** | 29.58mph | 47.6 kph |
| **Tire size** | 27-R49 (48 PR) E-4 | |
| **Wheelbase** | 180.06in | 4,570 mm |
| **Length** | 426.31in | 10,820 mm |
| **Maximum operating weight** | 351,344.70lb | 159,340 kg |
| **Payload** | 100.31t | 91 MT |

CC 1448

# Articulated Dump Trucks

# Articulated Dump Trucks

*The size, capacity, and shape of ADTs vary but the success of the design is based on articulation and maneuverability.*

**Articulated dump trucks (ADTs) are trucks that have an articulation joint and an oscillation ring between the cab and the body. The oscillating ring allows the cab section of the truck to move independently of the cargo body. When the truck is crossing uneven ground, the oscillation ensures that all wheels stay in contact with the ground to facilitate maximum traction as well as stability and maneuverability. The articulation joint allows the truck to bend in the middle and thereby serves as a steering function. The first ADTs were developed in Europe in the 1960s where difficult weather and often confined working conditions dictated the need for machines capable of carting large loads but also of being able to operate where the more traditional rigid-frame dump truck could not. The idea is essentially a development of the tractor and trailer combination but has grown into one of the most versatile pieces of earthmoving equipment available.**

*This Komatsu Moxy MT27, like all ADTs, articulates behind the operator cab.*

One of the first articulated dump trucks was the 4x2 machine manufactured by Northfield Engineering and displayed at the 1961

International Construction Equipment Exhibition at Crystal Palace, London. It was front-wheel drive and had a payload of 12.32 tons (11.18 tonnes), power-assisted steering, and had a turning circle of 18 feet 9 inches (5.72 meters). The uses to which ADTs can be put is almost endless and the machines are now to be found in mineral extraction, landfill, construction, and raw materials handling.

The major advantages of the ADT are perceived to be: maneuverability, versatility, ease of operation, productivity, and cost efficiency.

***A Terex 2566C ADT being utilized on a road construction scheme.***

**Maneuverability** An articulated truck's configuration allows it to turn up to 45 degrees to the right and left, giving the operator maneuverability in tight situations. The 6x6 drive capability, along with large off-road tires and automatic transmission, allows the ADT to continue working in situations that may well bog down other types of equipment.

**Versatility** The phenomenal growth in ADT use in recent years is testament to its versatility and practicality. The ADT, because of its abilities, can be used in areas where seasonal changes are extreme, and this has the advantage of extending the working season.

**Ease of operation** The ADTs are not complicated machines, and feature accessible control panels and gear systems. The cabs are designed to be comfortable for the operator to ensure productivity. Features such as automatic transmission leave the operator free to concentrate on the on-site tasks rather than shifting gears. Because of the numbers of ADTs in the contract hire market, their ease of operation is a bonus to those teaching and learning ADT operation.

**Productivity** The ADTs are productive under all working conditions and these machines feature the highest payload-to-weight ratio of any type of earthmoving equipment. Their cross-country ability means that specific routes need not be prepared, which is an important factor on job sites that will be short lived.

**Cost efficiency** The initial purchase price of an ADT is lower than other earthmoving equipment because fuel efficiency is high, and moving them between sites is less complex than much larger machinery because most will fit on a standard low-boy truck and do not require special movement permits.

The future for articulated dump trucks looks bright as a result of these factors and they are becoming widely accepted as the best way of moving quantities of material. There are numerous manufacturers of ADT-type trucks and a representative selection is contained within this chapter. In the specification panels, the measurement described as wheelbase is the distance between the front and second axles.

1

2

3

4

*1 The Aveling Barford RXD 28 tips in 16 seconds.*

*2 ZF transmission provides 6 forward and 3 reverse gears with lock up in all gears.*

*3 Each wheel is fitted with disc brakes and a spring actuated disc parking brake.*

*4 The rear suspension gives high stability, maintaining high traction through greater ground contact.*

# Aveling Barford ADTs

**English company Aveling Barford manufacture two models of ADT, the RXD 24 and RXD 28. The machines are assembled through use of components bought in by the company including Cummins diesel engines, ZF transmissions, and ZF axles. The models have varying payloads and rear-body capacities, both sizes of which are emptied through being raised by a six-stage hydraulic ram. The load-carrying body is made from wear-resistant steel and profiled to ensure clean load ejection. The rear suspension is the proven and uncomplicated rocking-beam arrangement, while the front relies on a self-levelling air system with hydraulic dampers. Disc brakes are fitted to all six hubs and the coupling for the articulating joint runs in cast nylon bearings.**

*The dump body is made from hard wearing steel and designed for clean load ejection. Steering is of the full orbital type.*

| AVELING BARFORD RXD 24 ARTICULATED DUMP TRUCK | |
|---|---|
| **Engine** | |
| Cummins LT10 | |
| **Engine power** | |
| 205 kW | |
| **Bore and stroke** | |
| n/a | |
| **Transmission** | |
| 6F, 3R | |
| **Maximum speed** | |
| 32.93mph | 53 kph |
| **Tire size** | |
| 20.50-R25 | |
| **Wheelbase** | |
| 159.57in | 4,050 mm |
| **Length** | |
| 383.56in | 9,735 mm |
| **Maximum operating weight** | |
| 78,698.66lb | 35,691 kg |
| **Payload** | |
| 24.25t | 22 MT |

*To maximize axle articulation the RXD 28 uses self compensating air suspension for the front axle and rocking beams for the rear axles.*

| AVELING BARFORD RXD 28 ARTICULATED DUMP TRUCK | |
|---|---|
| **Engine** | |
| Cummins LT10 | |
| **Engine power** | |
| 205 kW | |
| **Bore and stroke** | |
| n/a | |
| **Transmission** | |
| 6F, 3R | |
| **Maximum speed** | |
| 29.83mph | 48 kph |
| **Tire size** | |
| 23.50-R25 | |
| **Wheelbase** | |
| 159.57in | 4,050 mm |
| **Length** | |
| 383.56in | 9,735 mm |
| **Maximum operating weight** | |
| 93,617.69lb | 42,457 kg |
| **Payload** | |
| 27.56t | 25 MT |

# Bell ADTs

***The Bell 20B 6x4 Super Truck has two air bags controlled by a levelling valve which maintains ride height regardless of the load.***

**Bell Equipment are a South African company who manufacture a large range of articulated dump trucks. As well as differing payloads, they make ADTs with a variety of drive configurations: 4x2, 6x2, 6x4, and 6x6 machines are all listed in the company's range (the first digit refers to the number of wheels that the machine has and the second to the number of driven wheels). The machine designated B16B is available in both 4x2 and 6x2 forms, the B17B and B20B as 6x4 machines, the B25B as both 6x4 and 6x6, and the larger B30B, B35B, and B40B as 6x6. In the UK some of these machines are marketed as Heathfield Haulamatic ADTs, including the B25B.**

## Bell 16B ADT

In both six- and four-wheeled versions of the B16B the engine is an in-line, six-cylinder, turbocharged diesel engine and a ZF five-speed transmission. The front driven axle is a Bell 14T type, while trailing axles are 10T and 25T on the 6x2 and 4x2 models respectively. Front suspension is a leading-arm rigid subframe linked by pins to the chassis in order to allow vertical movement of the axle. Rear suspension is the proven walking beams system and total oscillation is unlimited. The range of transmission options for vehicles of

***Heavy duty shock absorbers are fitted to the B16B 6x2 to dampen suspension movement.***

similar size and payload allows the customer to choose either all-wheel drive, four wheel drive, or two wheel drive machines. The choice will depend on economic factors as well as taking in to consideration the type of terrain in which the machine is to be utilized. An ADT used primarily on surfaced or hard roads will not require all-wheel drive while if used on softer or more uneven ground, it may well do so. The extra driven axles are an additional cost in the construction of each ADT. The variety of ADTs both reflects the variety of uses to which the versatile machines are put and further increases the versatility of the vehicle type.

**BELL EQUIPMENT B16B 6X2 ARTICULATED DUMP TRUCK**

| | | |
|---|---|---|
| **Engine** | | |
| | ADE 366T | |
| **Engine power** | | |
| | 120 kW (160 hp) @ 2300 rpm | |
| **Bore and stroke** | | |
| | n/a | |
| **Transmission** | | |
| | 5F, 3R | |
| **Maximum speed** | | |
| | 29.83mph | 48 kph |
| **Tire size** | | |
| | 20.50-R25 | |
| **Wheelbase** | | |
| | 168.24in | 4,270 mm |
| **Length** | | |
| | 340.02in | 8,630 mm |
| **Maximum operating weight** | | |
| | 66,458.70lb | 30,140 kg |
| **Payload** | | |
| | 19.84t | 18 MT |

***The B35B 6x6 is one of the largest ADTs that Bell produces.***

***The Bell B25B 6x6 articulated dump truck.***

| BELL EQUIPMENT B25B 6X6 ARTICULATED DUMP TRUCK | |
|---|---|
| **Engine** | |
| ADE 366TI | |
| **Engine power** | |
| 168 kW (225 hp) @ 2300 rpm | |
| **Bore and stroke** | |
| n/a | |
| **Transmission** | |
| 6F, 3R | |
| **Maximum speed** | |
| 30.14mph | 48.5 kph |
| **Tire size** | |
| 23.50-R25 | |
| **Wheelbase** | |
| 162.80in | 4,132 mm |
| **Length** | |
| 356.92in | 9,059 mm |
| **Maximum operating weight** | |
| 86,250.78lb | 39,116 kg |
| **Payload** | |
| 25.35t | 23 MT |

| BELL EQUIPMENT B35B 6X6 ARTICULATED DUMP TRUCK | |
|---|---|
| **Engine** | |
| ADE 442T (Mercedes Benz) | |
| **Engine power** | |
| 260 kW (349 hp) @ 2100 rpm | |
| **Bore and stroke** | |
| n/a | |
| **Transmission** | |
| 5F, 2R | |
| **Maximum speed** | |
| 31.69mph | 51 kph |
| **Tire size** | |
| 26.50-R25 | |
| **Wheelbase** | |
| 176.28in | 4,474 mm |
| **Length** | |
| 415.04in | 10,534 mm |
| **Maximum operating weight** | |
| 70,339.50lb | 31,900 kg |
| **Payload** | |
| 62.83t | 57 MT |

***The Caterpillar D250E is the smallest 6x6 ADT manufactured by the company.***

# Caterpillar ADTs

**Caterpillar manufacture a total of six different articulated dump trucks; four are 6x6 machines and the remaining two 4x4. The smallest 6x6 model is the D250E and the largest the D400E. These machines have load capacities of 25.02 and 40.01 tons (22.70 and 36.30 tonnes) respectively.**

### Caterpillar D250E

The D250E features a body designed with a low center of gravity and a large body volume, which gives stability and allows loading in a variety of ways. For low-density materials, such as waste or coal, specially larger-capacity bodies are available, as are flatbeds and specialist conversions for specific jobs. Other options include an underslung tailgate for further improved material retention. Like many other Caterpillar products, the ADT features a sound-suppressed ROPS cab with such refinements as air conditioning, pressurized air, tinted laminated windshield, and an ergonomically designed dashboard all fitted as standard. The steering wheel is on a tiltable column and the operator's seat is of the air-suspension type.

The engine displaces 2.77 gallons (10.50 liters) and is a turbocharged, aftercooled in-line six with a direct-injection fuel system. Inside the cast

**CATERPILLAR D250E ARTICULATED TRUCK**

**Engine**
Caterpillar 3306

**Engine power**
205 kW (275 hp) @ 2200 rpm

**Bore and stroke**
4.77 x 5.99in — 121 x 152 mm

**Transmission**
5F, 2R

**Maximum speed**
50.9 kph

**Tire size**
23.50-R25

**Wheelbase**
148.14in — 3,760 mm

**Length**
391.52in — 9,937 mm

**Maximum operating weight**
96,314.40lb — 43,680 kg

**Payload**
25.02t — 22.7 MT

***The D400E is the largest capacity 6x6 ADT in the range. It has a rated payload of 40 tons (36.3 tonnes) and a body capacity of 28.6 cubic yards (21.8 cubic meters).***

block are alloy pistons, steel connecting rods, and a forged crankshaft. The five-speed transmission also features two reverse gears and shifting is electronically controlled. The three driven axles are similar to ensure commonality between parts. The front axle is mounted on a swinging cradle subframe arrangement, which pivots on the front frame, while the pair of rear axles is mounted on a beam assembly that allows all four rear wheels to remain in contact with the ground.

**Caterpillar D400E** The D400E is a considerably larger version of the D250E and, although it is similar in many ways, including the type of chassis frames around which it is based and its use of one of Caterpillar's range of diesel engines, it is different in others including the suspension arrangement. The front axle has hydraulic suspension while the rear axles feature what the manufacturer describes as a "Hydroflex" system. In this case the axle movement is hydraulically damped and load-transfer controlled. Two interconnected pairs of suspension cylinders work to ensure equal axle loading and ground contact, while allowing axle oscillation and wheel

# Komatsu Moxy

**The origins of Moxy articulated dump trucks are in the 1960s when a Norwegian, Birger Hatlebakk, started to make an advanced off-road vehicle based on an articulating design which he believed would make earthmoving operations easier in rough terrain. The first experimental models required development and in 1973 Hatlebakk's company Glamox, acquired the rights to produce another truck in production by another Norwegian company, Overaasen. Progress after this was fast and lead to the production of the D15, D16, and D16B articulated dump trucks. The company was owned by Glamox but was known as Moxy A/S and initially the machines were made for the Norwegian domestic market. However, as the concept of articulated dump trucks spread, Moxy began exporting. Dealerships around Europe were established and the European market grew through the seventies and eighties. In the United States Moxy Trucks of America Inc. was established by Leo H. Gerbus in Cincinnati, Ohio. Gerbus had 30 years of experience in the construction industry. In 1986 Moxy entered into an agreement with Komatsu Ltd. of Japan, the world's second-largest manufacturer of construction equipment, which allowed Moxy Trucks to be sold badged as Komatsu in markets that are not served by Moxy.**

In June 1991 the company was restructured and became partially owned by the Norwegian mining group A/S Olivin and partially owned by Komatsu Ltd. The company then began development of a new range of ADTs, including use of proprietary components such as engines from Scania and transmissions from ZF. Current Moxy trucks are constant six-wheel drive (6x6) articulated dump trucks in the 33.60–44.8-ton (30.48–40.64 tonne) capacity range. The trucks are intended for off-road

| MOXY MT30S ARTICULATED DUMP TRUCK | |
|---|---|
| **Engine** | |
| Scania DSC9 | |
| **Engine power** | |
| 187 kW @ 2200 rpm | |
| **Bore and stroke** | |
| n/a | |
| **Transmission** | |
| 6F, 3R | |
| **Maximum speed** | |
| 32.31mph | 52 kph |
| **Tire size** | |
| 23.50-R25 | |
| **Wheelbase** | |
| 116.19in | 2,949 mm |
| **Length** | |
| 156.89in | 3,982 mm |
| **Maximum operating weight** | |
| 99,225lb | 45,000 kg |
| **Payload** | |
| 30.09t | 27.3 MT |

use in tasks such as construction, land reclamation, and landfill. ADTs have earned a reputation for being easy to operate, economical, and productive even in the most difficult conditions.

### Moxy MT30 Articulated Dump Truck

The MT30LHS is a variant of the MT30 and is a multi-purpose, 33.60-ton (30.48-tonne) capacity, six-wheel drive, articulated dump truck which allows the transport of containers or roll-offs in rough terrain. The proven hook lift system accepts standard ISO containers but can also be fitted with water tanks, drilling platforms, and snow blowers meaning that the MT30LHS is a versatile machine. Like the MT30 it is powered by an electronically controlled automatic transmission and is driven by a turbocharged DSC9 Scania engine.

The MT30X is a 33.60-ton (30.48-tonne) capacity, six-wheel drive, articulated dump truck intended for landfill and construction applications. It features a permanent all wheel drive system with an automatic converter lock-up and torque proportioning differentials. It provides equal torque to all wheels without "wind up" in the transmission. Free-swinging rear tandem axle housings and Moxy's oscillation system keep the wheels in contact with the ground almost regardless of the ground conditions. The MT30X has a downward-sloping chassis and flat body floor, a feature unique to Moxy, which lowers the center of gravity and improves stability. A Scania DSC9 watercooled, turbocharged, six-cylinder diesel engine powers the MT30X and drives through an electronically controlled automatic ZF transmission.

### Moxy MT40 ADT

The MT40 is the largest truck in the Moxy line. It is a 44.80-ton (40.64-tonne) capacity, six-wheel drive ADT for moving large amounts of materials. The MT40 is powered by a V8 diesel engine and, like the smaller Moxy ADTs, features an electronically controlled automatic transmission.

| MOXY MT40 ARTICULATED DUMP TRUCK | |
|---|---|
| **Engine** | |
| Scania DS14 | |
| **Engine power** | |
| 304 kW (419 hp) @ 2100 rpm | |
| **Bore and stroke** | |
| n/a | |
| **Transmission** | |
| 6F, 3R | |
| **Maximum speed** | |
| 29.21 mph | 47 kph |
| **Tire size** | |
| 26.50-R25 | |
| **Wheelbase** | |
| 171.19in | 4,345 mm |
| **Length** | |
| 389.78in | 9,893 mm |
| **Maximum operating weight** | |
| 138,033lb | 62,600 kg |
| **Payload** | |
| 44.09t | 40 MT |

***The Moxy MT40 articulated dump truck can perform in virtually any conditions.***

***A fully laden Moxy articulated dump truck on a highway construction site.***

# Moxy ADTs at Work

The type of construction where ADTs are ideal are illustrated by a number of examples of jobs where Moxy machines are in operation. In order to lessen the threat of flooding in residential areas of Harris County, Texas, the Harris County Flood Control Department decided to widen 4.5 miles (7.24 kilometers) of the White Oak Bayou in North West Houston. While this work was being carried out, three detention basins would also be constructed. The county engineers divided the whole scheme into four phases, to be carried out over a 30-month period. A total of 1.5 million cubic yards (1.15 million cubic meters) of earth would have to be excavated to complete the job, and in some places these excavations would reach a depth of 40 feet (12.19 meters). Lloyd Engineering & Construction–Lecon Inc.–examined the sites and type of material to be moved before deciding on the use of excavators and ADTs as being the most feasible and economical way to carry out the work.

Ten Moxy MT30 ADTs supplied by the Houston-based company Con-Equip Inc. were put on the job, teamed up with Link-Belt excavators. The 6x6 capability of the Moxy trucks enabled them to carry full loads despite the wet and spongy conditions along the bayou's bank. Even during periods of rain when a bayou-side site becomes particularly wet, the fully laden ADTs were capable of handling the conditions and the washed-out site access roads.

Another construction company in the United States that relies on Moxy ADTs is the Hall Irwin Construction Company of Greeley, Colorado, who specialize in laying pipelines, water systems, and cross-country transmission lines. The company has a sand and gravel division which uses Moxy MT30 articulated dump trucks, supplied by Power Motive of Denver, Colorado, loaded by a Komatsu PC400LC-6 excavator. In Greeley the trucks move 6,720 tons (6,096 tonnes) of gravel per day whereas in another pit, because of shorter haul lengths, two of the Moxy ADTs can move the same amount.

In New Jersey, at the Edgeboro Landfill in East Brunswick, approximately one million cubic yards (764,600.00 cubic meters) of municipal waste needed moving from one area of the landfill to another. Edgeboro Disposal Inc. determined that 45 acres (18.21 hectares) of refuse that was between 20 and 30 years old had to be moved in an eight-month period. Brunswick Services Co. Inc. took on the job conscious that work during the winter months on top of varying degrees of thawing refuse could be problematic for on-site equipment. The Binder Machinery Company of South Plainfield, New Jersey, supplied 25 pieces of equipment for the task, including seven Moxy MT40 articulated dump trucks as well as excavators and dozers. The Moxy MT40 ADTs were equipped with optional high-volume tail gates designed to carry loads of lighter and sloppy material. Site preparation began in October and was followed by the excavation phase. Brunswick Services Inc. built 3,500 feet (1,066.80 meters) of haul roads across the site including a landfill road. Through the duration of the job, between 6,000 and 9,000 cubic yards (4,587.60 and 6,881.40 cubic meters) of material would be excavated per day depending on weather and daylight conditions. The alternate winter freezing and thawing created tough conditions on site and backing the trucks onto soft garbage was a delicate task. The 6x6 capability of the ADTs ensured cycle times were kept to a minimum.

# Terex ADTs

***A Terex 2566C articulated dump truck dumping a load of gravel during road construction. The rear body has a heaped capacity of 17.6 cubic yards (13.5 cubic meters).***

**Terex are a long-established manufacturer of earthmoving equipment but one of their first articulated dump trucks was the B20 of the late 1960s. It was a bottom-discharge unit capable of 28 miles per hour (45.08 kilometers per hour) and of carrying 13 cubic yards (9.94 cubic meters) struck. The bottom dump doors were opened by a cable and winch arrangement. Currently Terex manufacture a range of articulated dump trucks with varying payloads from 25.35 to 40.23 tons (23 to 36.5 tonnes). The machines' designations start with the metric tonnage so that the 28-ton (25.40-tonne) machine is the 2566C, while the 30.24-ton (27.43-tonne) model is the 2766C and so on.**

The Terex machines are six-wheel drive and designed to move a variety of loads over widely varying ground conditions. The six-wheel drive and comprehensive drivetrain preserve traction on slippery surfaces or in loose material and the suspension accommodates uneven ground as does the interframe oscillation.

**Terex 2566C Articulated Truck**

The Terex 2566C ADT is based around all-welded high-grade steel frames of a rectangular box-section design. The frames are engineered to exceed the stresses imposed when used on uneven sites. The front frame accommodates the engine, transmission, hydraulic and fuel tanks, and supports the cab. Steering articulation is by two widely spaced vertical pivot pins fitted in sealed taper roller bearings. The front suspension is intended to be maintenance free; a leading arm subframe carries the front axle and pivots on the frame. Suspension is by air-filled rubber bellows which maintain ride height regardless of load. Hydraulic dampers are also fitted to control movement on both bump and rebound. The rear suspension consists of axles linked to the chassis via three rubber bushed links which provide both longitudinal location and torque reaction control. Lateral location is by means of transverse links fitted with spherical bearings. This design ensures that the inter-axle drive-line angles are maintained at the optimum in all axle positions, ensuring longevity of the components. The loads on the rear axles are balanced by the centrally pivoting equalizer beams. The axles are fitted with limited-slip differentials which engage automatically as conditions dictate. The center axle has two differentials, one to drive the half shafts for the normal cross-axle drive and the other for through differential drive. All the three axles have single reduction, spiral bevel gear differentials and a secondary reduction through outboard mounted planetary gears. Air/oil actuated disc brakes are fitted to each hub. The whole machine is powered by a C Series Cummins diesel engine and a power shift ZF 6WG 200 transmission with integral torque converter. Gear shifts are operated by electric control of hydraulically operated multiplate clutches. Automatic converter lock-up engages in all forward gears to eliminate slippage losses.

| TEREX 2566C ARTICULATED TRUCK | |
|---|---|
| **Engine** | |
| Cummins 6CTA8.3-C | |
| **Engine power** | |
| 190 kW (255 hp) @ 2000 rpm | |
| **Bore and stroke** | |
| 4.49 x 5.32in | 114 x 135 mm |
| **Transmission** | |
| 6F, 3R | |
| **Maximum speed** | |
| 26.10mph | 42 kph |
| **Tire size** | |
| 20.50-R25 | |
| **Wheelbase** | |
| 160.75in | 4,080 mm |
| **Length** | |
| 372.13in | 9,445 mm |
| **Maximum operating weight** | |
| 92,257.20lb | 41,840 kg |
| **Payload** | |
| 28t | 25.35 MT |

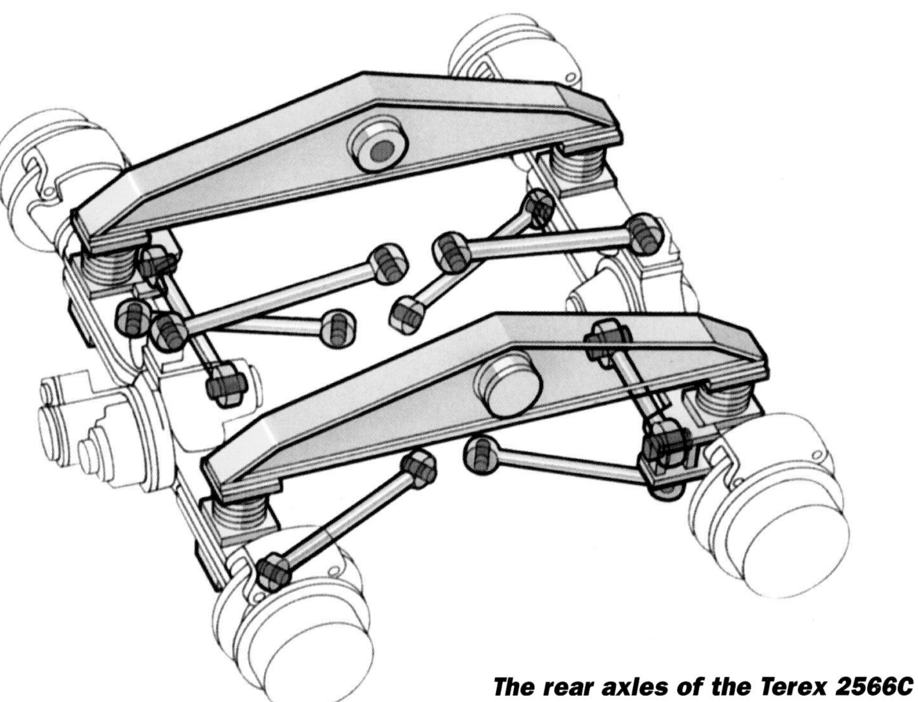

***The rear axles of the Terex 2566C are connected to the chassis by three rubber-bushed links which control torque reactions and locate the axles which are mounted onto a pair of centrally pivoting equalizer beams. Rubber bushings are shown in red.***

| TEREX 2766C ARTICULATED TRUCK | |
|---|---|
| **Engine** | |
| Cummins 6CTA8.3-C | |
| **Engine power** | |
| 190 kW (255 hp) @ 2000 rpm | |
| **Bore and stroke** | |
| 4.49 x 5.32in | 114 x 135 mm |
| **Transmission** | |
| 6F, 3R | |
| **Maximum speed** | |
| 27.96mph | 45 kph |
| **Tire size** | |
| 23.50-R25 | |
| **Wheelbase** | |
| 160.75in | 4,080 mm |
| **Length** | |
| 372.13in | 9,445 mm |
| **Maximum operating weight** | |
| 98,431.20lb | 44,640 kg |
| **Payload** | |
| 30.24t | 27.43 MT |

***The Terex 2766C articulated truck is six wheel drive and with its 255 hp (190 kW) engine has a payload of 27 tons (25 tonnes).***

| TEREX 3066C ARTICULATED TRUCK | |
|---|---|
| **Engine** | |
| Cummins MTA11-C300 | |
| **Engine power** | |
| 224 kW (300 hp) @ 2000 rpm | |
| **Bore and stroke** | |
| 4.93 x 5.79in | 125 x 147 mm |
| **Transmission** | |
| 6F, 3R | |
| **Maximum speed** | |
| 27.96mph | 45 kph |
| **Tire size** | |
| 23.50-R25 | |
| **Wheelbase** | |
| 160.75in | 4,080 mm |
| **Length** | |
| 372.13in | 9,445 mm |
| **Maximum operating weight** | |
| 104,858.78lb | 47,555 kg |
| **Payload** | |
| 30.24t | 27.43 MT |

***The Terex 3066C is also a 6x6 but features a larger engine that produces 300 hp (224 kW) to enable it to move 30 tons (27 tonnes).***

**Terex 4066C Articulated Truck**

While the 2766C and 3066C can be considered as larger versions of the 2566C through the use of similar engines and the fact that the trucks have similar dimensions, the same cannot be said of the 4066C. This articulated truck is a much bigger machine with a considerably larger payload and larger dimensions. The rear body has a heaped capacity of 28.78 cubic yards (22.00 cubic meters). The Detroit Diesel Series 60 engine used is an in-line six-cylinder, four-cycle diesel, turbocharged, water-cooled unit with electronic engine management known as Detroit Diesel Electronic Controls (DDEC).

***The 400 hp (298 kW) Detroit Diesel engine enables the 4066C ADT to move 40 tons (36.5 tonnes) in the rear body which has a heaped capacity of 29 cubic yards (22 cubic meters).***

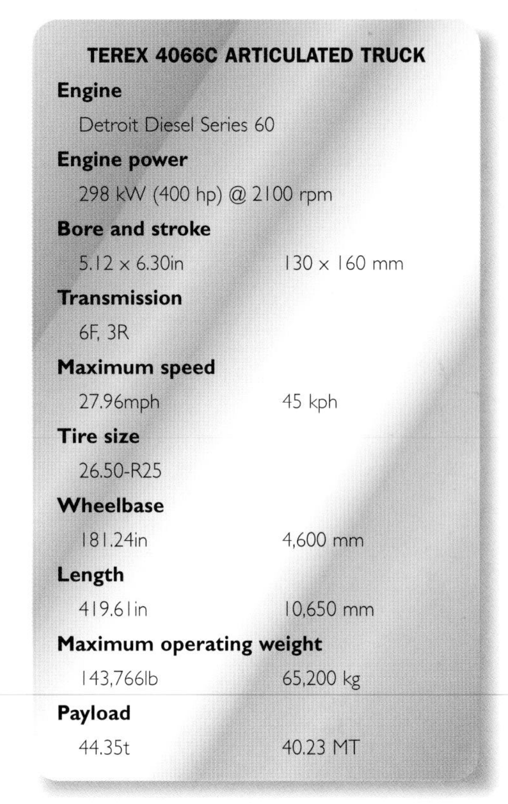

| TEREX 4066C ARTICULATED TRUCK | |
|---|---|
| **Engine** | |
| Detroit Diesel Series 60 | |
| **Engine power** | |
| 298 kW (400 hp) @ 2100 rpm | |
| **Bore and stroke** | |
| 5.12 x 6.30in | 130 x 160 mm |
| **Transmission** | |
| 6F, 3R | |
| **Maximum speed** | |
| 27.96mph | 45 kph |
| **Tire size** | |
| 26.50-R25 | |
| **Wheelbase** | |
| 181.24in | 4,600 mm |
| **Length** | |
| 419.61in | 10,650 mm |
| **Maximum operating weight** | |
| 143,766lb | 65,200 kg |
| **Payload** | |
| 44.35t | 40.23 MT |

***The Terex 4066C is powered by the Detroit Diesel Series 60 engine while the smaller Terex ADTs are driven by Cummins units.***

# Volvo BM Articulated Haulers

***The Volvo A25C is a 6x6 ADT and one of a range of five made by the company. Volvo BM also manufacture a smaller 4x4 articulated hauler.***

**Volvo Construction Equipment are the world's largest manufacturer of articulated haulers. As well as this, they also manufacture ranges of wheel loaders, rigid haulers, and hydraulic excavators. The Swedish based company's products are marketed all around the world under the well-known names of Volvo BM, Michigan, Euclid, Zettelmeyer, and Åkerman. Volvo BM offer a range of six articulated haulers: the A20C 6x6, A25C 4x4, A25C 6x6, A30C 6x6, A35C 6x6, and the A40 6x6. Payloads vary from 24.80 tons (22.50 tonnes) to 39.68 tons (36 tonnes). In the former capacity range comes the A25C 6x6, of which there is also a 4x4 variant.**

Volvo BM have refined the concept of the articulated dump truck–although they term it hauler–and their machines feature all-wheel drive, 100 percent differential lock on all axles, and operator-friendly cabs. The automatic transmission is designed to be both fast and yet prolong drivetrain

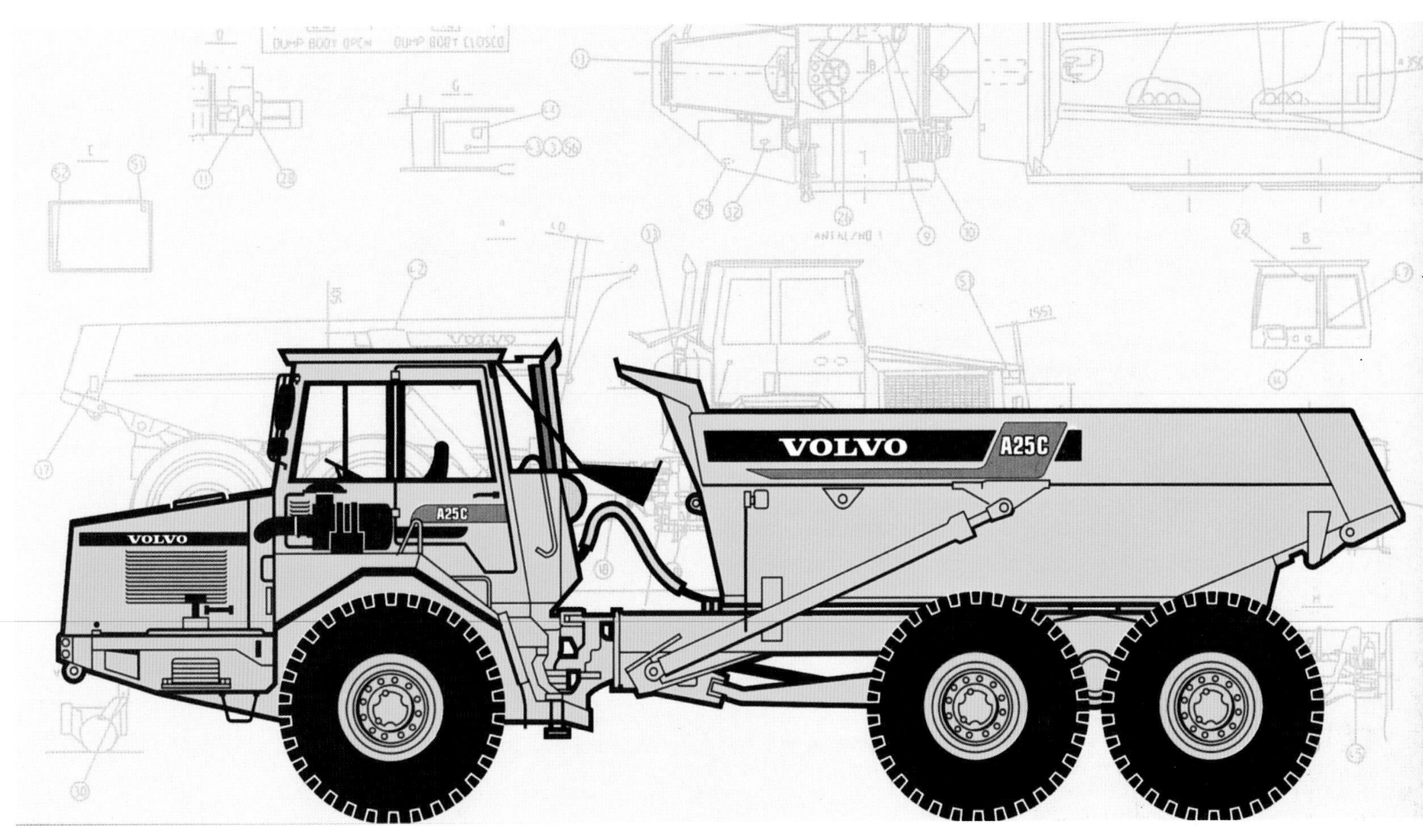


***The side elevation of the Volvo A25C articulated dump truck. Hydraulic ram and hoses are clearly evident.***

component wear. It is geared to maximize use of engine power and maintain fuel economy.

Perceived advantages of Volvo BM articulated haulers are their competent off-road ability, meaning that they can maintain cycle times despite operating in rough terrain or on sites without access roads. The articulated steering system and rotating frame joint keep the machine operable in different conditions and is the key to the Volvo's off-road performance. The three-point suspension design allows for high-speed load haulage on poor roads.

VOLVO

**VOLVO A25C 6X6 ARTICULATED HAULER**

**Engine**
Volvo TD73 KCE

**Engine power**
190 kW (255 hp) @ 2400 rpm

**Bore and stroke**
4.13 x 5.12in — 104.77 x 130 mm

**Transmission**
10F, 2R

**Maximum speed**
32.31mph — 52 kph

**Tire size**
23.50-R25

**Wheelbase**
164.10in — 4,165 mm

**Length**
381.20in — 9,675 mm

**Maximum operating weight**
88,795.35lb — 40,270 kg

**Payload**
24.80t — 22.5 MT

| VOLVO A35C 6X6 ARTICULATED HAULER | |
|---|---|
| **Engine** | |
| Volvo TD122 KAE | |
| **Engine power** | |
| 245 kW (328 hp) @ 2100 rpm | |
| **Bore and stroke** | |
| 5.12 x 5.91in | 130 x 150 mm |
| **Transmission** | |
| 12F, 3R | |
| **Maximum speed** | |
| 32.31mph | 52 kph |
| **Tire size** | |
| 26.50-R25 | |
| **Wheelbase** | |
| 177.10in | 4,495 mm |
| **Length** | |
| 429.30in | 10,898 mm |
| **Maximum operating weight** | |
| 127,228.50lb | 57,700 kg |
| **Payload** | |
| 35.27t | 32 MT |

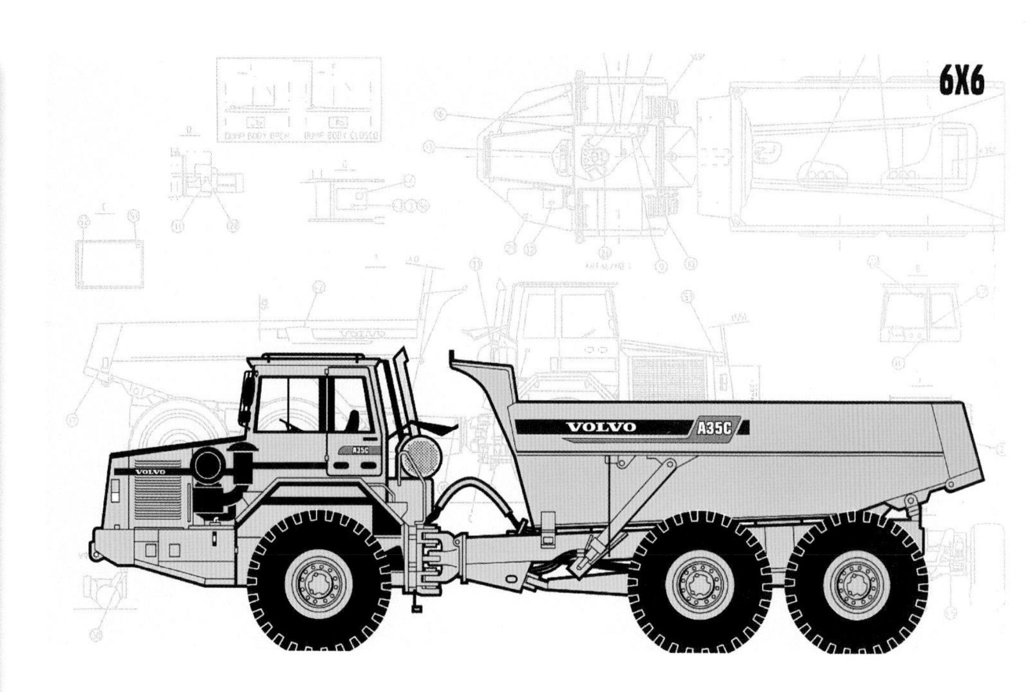


***The Volvo A35C articulated dump truck features three driven axles making it a 6x6 machine.***

| VOLVO A40 6X6 ARTICULATED HAULER | |
|---|---|
| **Engine** | |
| Volvo TD122 KFE | |
| **Engine power** | |
| 1297 kW (398 hp) @ 2100 rpm | |
| **Bore and stroke** | |
| 5.12 x 5.91in | 130 x 150 mm |
| **Transmission** | |
| 12F, 2R | |
| **Maximum speed** | |
| 32.69mph | 52.6 kph |
| **Tire size** | |
| 29.50-R25 | |
| **Wheelbase** | |
| 175.01in | 4,442 mm |
| **Length** | |
| 438.72in | 11,135 mm |
| **Maximum operating weight** | |
| 143,655.75lb | 65,150 kg |
| **Payload** | |
| 39.68t | 36 MT |

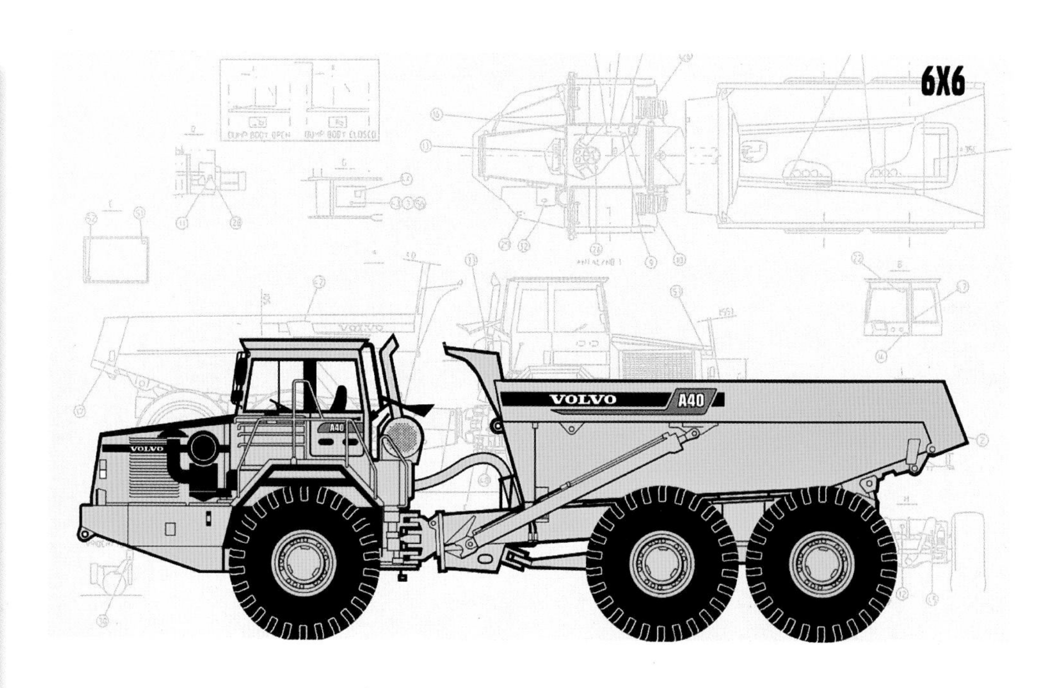


***The Volvo A40 is also a 6x6 and has the largest payload of the Volvo range.***

***Articulated dump trucks are now almost as common a sight on construction jobs as the ubiquitous bulldozer.***

D6H LGP
Alfred
McAlpine
Plant

# Bulldozers

# Bulldozers

*During World War Two the bulldozer proved its worth as a military machine. This picture shows an American unit with a bulldozer and several jeeps during the latter stages of the campaign in Europe.*

In the years after the First World War the Best Company continued their work with crawler tracked machinery. In 1921 they introduced a new machine, the Best 30 Tracklayer. This crawler was fitted with a light-duty bulldozer blade, was powered by an internal combustion engine, and had an enclosed cab. From the war through the 1920s there was a considerable amount of litigation involving patents and types of tracklayers. Two companies were frequently named in the litigation: Best and Holt. Holt's patent for tracklayers left him in a position to charge a licence fee to other manufacturers of the time, including Monarch, Bates, and Cletrac. The First World War intervened and much of Holt's production went to the United States Army, while Best supplied farmers. The two companies competed in all markets and neither had a significant advantage over the other. Eventually in 1925 two companies, Holt and Best, effectively merged to form The Caterpillar Tractor Company. In late 1925 the new Caterpillar Company published prices for its product line: the Model 60 sold for $6,050, the Model 30 for $3,665, and the 2.24-ton (2.03-tonne) for $1,975.

**The consolidation of the two brands into one company proved its value in the next few years; the prices of the big tracklayers were cut, the business increased, and sales more than doubled.**

***The shape of the bulldozer blade fitted varies depending upon the type of material to be dozed.***

Starting in 1931 all Caterpillar machinery left the factory painted Highway Yellow. This was seen as a way of brightening up the machines as an attempt to lift the gloom of the Depression and for safety reasons because machines increasingly being used in road construction had to be visible so that motorists would see them. It caught on slowly at first but eventually became the standard color for all construction equipment.

The diesel engine came in 1935 and model designations began RD—Rudolf Diesel's initials—and were finished with a number that related to the crawler's size and engine power, so there were RD8, RD7, and RD6 machines soon followed by the RD4 of 1936. Other stories claim that the "R" stands for Roosevelt, the "D" for Diesel, and the "8" equates approximately to the machine's engine capacity. By this time the United States Forest Service were using machines such as the Cletrac Forty, with an angled blade on the front, so Caterpillar built one with a LaPlante-Choate Trailblazer blade attached. Ralph Choate had started in business by building blades to be fastened to the front of other people's crawlers; his first one was used on road construction work between Cedar Rapids and Dubuque, Iowa.

American and British companies were leaders in bulldozer design. Initially they featured manually controlled blades but later incorporated electric control. During the 1920s Robert Gilmour Le Tourneau, an American contractor who manufactured earthmoving equipment for Holt, Best, and later Caterpillar prime movers, developed a new system of power control which began to widen the scope of the dozer. All three control systems featured winch and cable actuation until the development of hydraulically lifted and lowered blades. One of the first British machines to be so equipped was the Vickers Vigor developed from the Vickers VR-Series crawlers. Hydraulics were first used in the late 1930s in time for bulldozers to make a big impression during the Second World War. Bulldozers first came to military planners' attention after their use in removing beach defenses and even occupied pillboxes during the campaign at Guadalcanal where Aurelio Tassone received the Silver Star for removing a hostile pillbox under fire during a beach landing. He drove up the beach with the blade raised to shield himself from enemy fire then dropped it as he hit the pillbox. Later tanks would be equipped with bulldozer blades to assist in clearing obstacles. The Caterpillar D7 saw service in all theaters of operation during the Second World War, and General Eisenhower credited it as being one of the machines that won the war. In recent years the name "bulldozer" has become shortened so that in the vernacular of the day are usually referred to as dozers.

*Caterpillar's D5M uses an elevated sprocket which is designed to give better balance, performance, and undercarriage component life.*

# Caterpillar Dozers

*Dozers such as the D5M are based around a high strength steel mainframe on which provision is made to bolt the major mechanical components.*

**Caterpillar make a range of 26 tracked dozers, including low ground pressure (LGP) variants and three models of wheeled dozers, such is the popularity and versatility of the proven dozer concept. The operating weights of the D3C III and D11R—the smallest and largest in Caterpillar's tracked dozer range respectively—give an indication of the difference in size. The D3C III operates at 7.83 tons (7.1 tonnes), while the D11R operates at 108.47 tons (98.4 tonnes).**

**Caterpillar D5M Track-type Tractor**
In terms of operating weight, the D5M is found around the middle of Caterpillar's range and is available in both standard and low ground pressure applications. The low ground pressure versions feature considerably wider track shoes to reduce ground pressure.

The Caterpillar 3116 engine is of the direct-injection type, and turbocharged to improve response and performance at low to medium engine revolutions per minute. The engine is resiliently mounted for quieter operation and less vibration. To ensure longevity the engine is constructed around a one-piece castiron cylinder block topped with a one-piece cylinder head with replaceable stainless steel inlet valve seats and nickel alloy exhaust seats. The cylinders feature full-length water cooling for maximum heat transfer and the oil is cooled to

| CATERPILLAR D5M XL TRACK-TYPE TRACTOR | |
|---|---|
| **Engine** | |
| Caterpillar 3116 | |
| **Engine power** | |
| 90 kW (121 hp) @ 2100 rpm | |
| **Bore and stroke** | |
| 4.14 x 5.00in | 105 x 127 mm |
| **Transmission** | |
| 3F, 3R | |
| **Maximum speed** | |
| 6.17mph | 9.93 kph |
| **Track gauge** | |
| 69.74in | 1,770 mm |
| **Length of track on ground** | |
| 94.09in | 2,388 mm |
| **Length of basic tractor** | |
| 139.63in | 3,544 mm |
| **Maximum operating weight** | |
| 25,798.50lb | 11,700 kg |
| **Blade width** | |
| 121.23in | 3,077 mm |

maintain optimum oil temperature. The pistons are two piece with forged steel crowns for long life and the oil pump is mounted low on the engine for quick start-up lubrication. The engine is linked to a torque converter transmission designed to protect the drivetrain from shock loads. The transmission shifts through Caterpillar's proven powershift system which is of a planetary design and delivers fast, smooth changes, and perimeter mounted clutches provide both fast heat dissipation and a large contact area for long service life. Both clutches and brakes are oil cooled.

These major components are mounted to the steel mainframe which is designed to absorb high-impact shock loads and twisting. The mainframe is assembled through robotic welding to ensure consistent quality. The track system is designed so that the final drive and associated power train parts are raised above the work area to isolate them from a variety of loads including ground-induced impact, and implement and roller frame alignment which extends power-train component life. The high sprocket position keeps sprocket teeth, bushings, and final drives away from abrasive materials and moisture, again to ensure long-lasting component life. The low ground pressure undercarriage is designed to allow the D5M to operate in soft and spongy conditions. Lower ground pressure is achieved through use of wider track hoes, long track frame, and wider gauge to increase contact area and provide flotation in swampy conditions. A variety of single grouser shoe widths is available for different operating conditions. It should be borne in mind when considering the figures quoted in the specification panels that certain of the figures may vary if specific options are chosen, including length of the machine if additional equipment such as rippers are installed. Blade width can vary depending on customer choice and the task to be undertaken, and these choices will affect operating weights and speeds. Caterpillar work tools include a variety of designs of dozer blade, rippers (used for breaking ground), and winches. The operator cabs are as well appointed as in the rigid and articulated dump trucks.

***This dozer blade is what Cat describe as a VPAT–Variable Pitch Power Angle and Tilt–to make it suitable for various applications.***

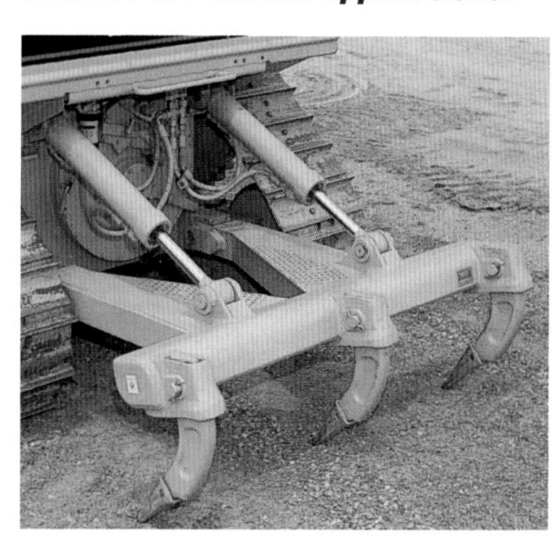

***The Ripper is another specialist tool that can be supplied with dozers such as the D5M. It is intended to break up the ground surface.***

***The Cat D10R has a 28.7 cubic yard (22 cubic meter) blade capacity and its Cat 3412E engine delivers 613 horsepower gross (457 kW).***

**Caterpillar D10R Track-type Tractor**
The D10R is one of the largest dozers in the range. The D10R is powered by a V12 diesel engine which is both turbocharged and aftercooled. The transmission utilizes a torque divider, a single stage torque converter with an output torque divider. This unit sends 75 percent of engine torque through the converter and the remaining 25 percent through a direct drive shaft to ensure drive-line efficiency and high torque multiplication. The planetary power- shift transmission has three forward and three reverse (3F,3R) speeds and uses large-diameter, oil-cooled clutches. Other parts of the D10R are similar to the smaller Caterpillar dozers, including the elevated drive sprockets.

***The undercarriage is designed with suspension in order to absorb impact loads and maintain maximum track contact with the ground.***

| CATERPILLAR D10R TRACK-TYPE TRACTOR | |
|---|---|
| **Engine** | |
| Caterpillar 3412E | |
| **Engine power** | |
| 457 kW (613 hp) @ 1900 rpm | |
| **Bore and stroke** | |
| 5.40 x 5.99in | 137 x 152 mm |
| **Transmission** | |
| 3F, 3R | |
| **Maximum speed** | |
| 7.77mph | 12.5 kph |
| **Track gauge** | |
| 100.47in | 2,550 mm |
| **Length of track on ground** | |
| 210.04in | 5,331 mm |
| **Length of basic tractor** | |
| 210.04in | 5,331 mm |
| **Maximum operating weight** | |
| 145,009.62lb | 65,764 kg |
| **Blade width** | |
| 191.56in | 4,862 mm |

**Caterpillar D11R Track-type Tractor**

The D11R is the biggest dozer in Caterpillar's comprehensive range, and its various blades have massive capacities. The 11SU blade is rated at 35.58 cubic yards (27.2 cubic meters), while the 11U blade is rated at 44.50 cubic yards (34.4 cubic meters). These have widths of 220.64 and 250.51 inches (5,600 and 6,358 mm) respectively, and both are 93.38 inches (2,370 mm) high. As with the smaller models, specialist tools are available including single and multi-shanked rippers. These are used for breaking up material and are designed for penetration and thoroughly ripping up a variety of materials.

***A D11R dozing, the elevated sprocket and single shank ripper are clearly visible.***

**CATERPILLAR D11R TRACK-TYPE TRACTOR**

| | | |
|---|---|---|
| **Engine** | | |
| Caterpillar 3508 | | |
| **Engine power** | | |
| 609 kW (817 hp) @ 1800 rpm | | |
| **Bore and stroke** | | |
| | 6.70 x 7.49in | 170 x 190 mm |
| **Transmission** | | |
| 3F, 3R | | |
| **Maximum speed** | | |
| | 7.21mph | 11.6 kph |
| **Track gauge** | | |
| | 114.10in | 2,896 mm |
| **Length of track on ground** | | |
| | 242.82in | 6,163 mm |
| **Length of basic tractor** | | |
| | 242.82in | 6,163 mm |
| **Maximum operating weight** | | |
| | 217,000.67lb | 98,413 kg |
| **Blade width** | | |
| | 220.64in | 5,600 mm |

***Specific tools have been designed to make full use of the power of the D11R including the 11SU blade (1) and the single shank ripper (2), a multi-shank ripper is also available.***

***The Caterpillar D6H LGP at work. The LGP suffix indicates a low ground pressure machine. Lower ground pressure is achieved through wider and longer tracks which spread the load.***

# Komatsu Dozers

**Komatsu are a long-established Japanese bulldozer manufacturer; their factory is in the Ishikawa Prefecture of Japan. Production of the company's D50 Series started in 1947 when the D50A1 was manufactured. It was a conventionally designed machine and was powered by the company's own 60-horsepower 4D120 diesel engine. By June 18, 1970 Komatsu had made 50,000 D50 machines. The current Komatsu range includes 23 dozers, including the world's largest—the Komatsu D575A-2 Super. This monster has an operating weight of 156.53 tons (142 tonnes), a blade capacity of between 44.47 and 88.94 cubic yards (34 and 68 cubic meters), and exerts a ground pressure of 2.0 kg/cm$^2$.**

**Komatsu D65E-12, D65EX-12, D65P-12, and D65PX-12 Bulldozers**

These are medium-class bulldozers in which maneuverability is optimized through use of wrist-controlled, single-lever steering and directional shifts. As can be seen from the specification panels, the D65EX-12 is simply a slightly higher-performance version of the D65E-12, and the D65PX-12 a higher-performance version of that. The former model is available in both standard track and long-track versions.

***Dozers, such as this D65EX, are capable of working on inclines and require a high degree of maneuverability for ease of operation.***

| KOMATSU D65EX-12 BULLDOZER | |
|---|---|
| **Engine** | |
| Komatsu S6D125E | |
| **Engine power** | |
| 142 kW (190 hp) @ 1950 rpm | |
| **Bore and stroke** | |
| 4.93 x 5.91in | 125 x 150 mm |
| **Transmission** | |
| 3F, 3R | |
| **Maximum speed** | |
| 6.59mph | 10.6 kph |
| **Track gauge** | |
| 80.77in | 2,050 mm |
| **Length of track on ground** | |
| 129.43in | 3,285 mm |
| **Length of basic tractor** | |
| 174.35in | 4,425 mm |
| **Maximum operating weight** | |
| 40,957.88lb | 18,575 kg |
| **Blade width** | |
| 136.32in | 3,460 mm |

***The D65EX-12 has good maneuverability in muddy terrain.***

| KOMATSU D65PX-12 BULLDOZER | |
|---|---|
| **Engine** | |
| Komatsu S6D125E | |
| **Engine power** | |
| 142 kW (190 hp) @ 1950 rpm | |
| **Bore and stroke** | |
| 4.93 x 5.91in | 125 x 150 mm |
| **Transmission** | |
| 3F, 3R | |
| **Maximum speed** | |
| 6.59mph | 10.6 kph |
| **Track gauge** | |
| 74.07in | 1,880 mm |
| **Length of track on ground** | |
| 29.27yd | 26,755 mm |
| **Length of basic tractor** | |
| 171.98in | 4,365 mm |
| **Maximum operating weight** | |
| 43,251.08lb | 19,615 kg |
| **Blade width** | |
| 156.42in | 3,970 mm |

***This D65PX-12 is equipped with the ROPS-roll over protection system-cab. Its minimum turning radius is 2.96 yards (2.7 meters).***

In semi-U tilt dozer format with a steel cab and a ROPS canopy, standard equipment and operator on board, the ground pressure exerted by the four machines is as follows: D65E-12, 0.67 kg/cm$^2$; D65EX-12 (standard track), 0.68 kg/cm$^2$; D65EX-12 (long track), 0.57 kg/cm$^2$; and D65PX-12,

0.31 kg/cm$^2$. Three differing dozer blades are available for these Komatsu models—namely semi-U tilt dozer, straight tilt dozer, and angle dozer—and choice depends on the use to which the machine will be put. One aspect of the operation of these machines where they vary considerably is in turning circle, which depends on the type of steering system fitted. Minimum turning radii of the four machines are as follows: D65E-12, 10.50 feet (3.2 meters); D65EX-12, 7.22 feet (2.2 meters); D65P-12, 11.81 feet (3.6 meters); and D65PX-12, 8.86 feet (2.7 meters). The X-suffixed models feature hydrostatic steering systems (HSS), while the others are turned by wet, multiple-disc steering clutches.

**Komatsu D155A-3 Bulldozer**

The D155A-3 is a refined version of the earlier D155A-2. Improvements include a Komatsu-designed resilient equalized undercarriage (REU) and a single-lever joystick for speed, steering, and directional changes. The REU features X-shaped track bogies that independently seesaw to enable the machine to maximize traction on uneven ground, as the three bogies on each track automatically follow the contours of the ground. In order to decrease vibration and shock, rubber shock absorbers are mounted on the bogies.

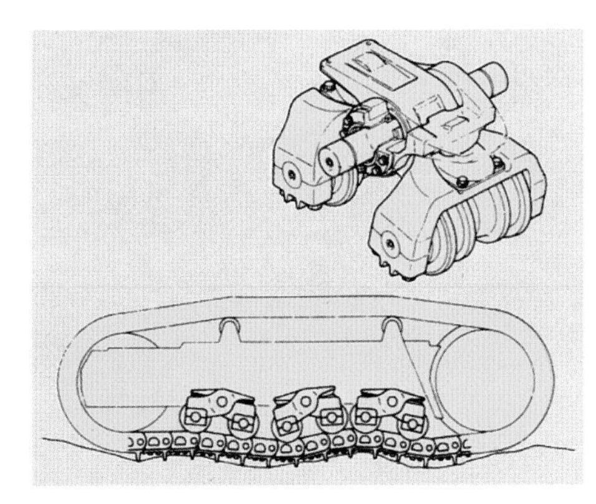

***The D155A-3 is fitted with what Komatsu call a Resilient Equalized Undercarriage (REU) where the bogie move independently to maximize track ground contact.***

| KOMATSU D155A-3 BULLDOZER | |
|---|---|
| **Engine** | |
| Komatsu S6D140 | |
| **Engine power** | |
| 225 kW (302 hp) @ 1900 rpm | |
| **Bore and stroke** | |
| 5.52 x 6.50in | 140 x 165 mm |
| **Transmission** | |
| 3F, 3R | |
| **Maximum speed** | |
| 6.84mph | 11 kph |
| **Track gauge** | |
| 89.04in | 2,260 mm |
| **Length of track on ground** | |
| 126.47in | 3,210 mm |
| **Length of basic tractor** | |
| 224.58in | 5,700 mm |
| **Maximum operating weight** | |
| 85,333.50lb | 38,700 kg |
| **Blade width** | |
| 155.83in | 3,955 mm |

***The Komatsu D155A-3 bulldozer is equipped with a semi-U dozer blade and multi-shank ripper.***

***The D275A-2 dozer seen here at work is powered by a 410 hp (301 kW) diesel engine. It is equipped with a dozer blade and single shank ripper.***

| KOMATSU D275A-2 BULLDOZER | |
|---|---|
| **Engine** | |
| Komatsu S6D170 | |
| **Engine power** | |
| 301 kW (405 hp) @ 1800 rpm | |
| **Bore and stroke** | |
| 6.70 x 6.70in | 170 x 170 mm |
| **Transmission** | |
| 3F, 3R | |
| **Maximum speed** | |
| 7.33mph | 11.8 kph |
| **Track gauge** | |
| n/a | |
| **Length of track on ground** | |
| 135.93in | 3,450 mm |
| **Length of basic tractor** | |
| 242.31in | 6,150 mm |
| **Maximum operating weight** | |
| 107,515.80lb | 48,760 kg |
| **Blade width** | |
| 169.42in | 4,300 mm |

**Komatsu D275A-2 Bulldozer**

The assets of the D275A-2 are seen as its productivity; its big diesel engine has the highest output in its class and it is the heaviest machine in its class—the 30.08-cubic yard (23-cubic meter) class. It has a conventional undercarriage drive design and a dual-tilt design of blade to ensure productivity. The machine is equipped with electronic monitoring systems to minimize down time, and its major components are modular to enable power-train removal without oil spillage and to allow easy replacement.

**Komatsu D375A-2 Bulldozer**

This Komatsu dozer shares many of its components with the smaller D275A-2 machine, including much of the undercarriage. It has a higher capacity than the smaller model however. The blade capacity in semi-U dozer form is 22.63 cubic yards (17.3 cubic meters) and in U dozer form is 27.34 cubic yards (20.9 cubic meters). The dual-tilt dozer option reduces operator effort, while increasing productivity, because it allows greater flexibility in setting the blade cutting angle for all types of material and inclination thereby making the dozer more versatile.

| KOMATSU D375A-2 BULLDOZER | |
|---|---|
| **Engine** | |
| Komatsu SA6D170 | |
| **Engine power** | |
| 391 kW (525 hp) @ 1800 rpm | |
| **Bore and stroke** | |
| 6.70 x 6.70in | 170 x 170 mm |
| **Transmission** | |
| 3F, 3R | |
| **Maximum speed** | |
| 7.33mph | 11.8 kph |
| **Track gauge** | |
| n/a | |
| **Length of track on ground** | |
| 150.31in | 3,815 mm |
| **Length of basic tractor** | |
| 323.08in | 8,200 mm |
| **Maximum operating weight** | |
| 139,444.20lb | 63,240 kg |
| **Blade width** | |
| 184.98in | 4,695 mm |

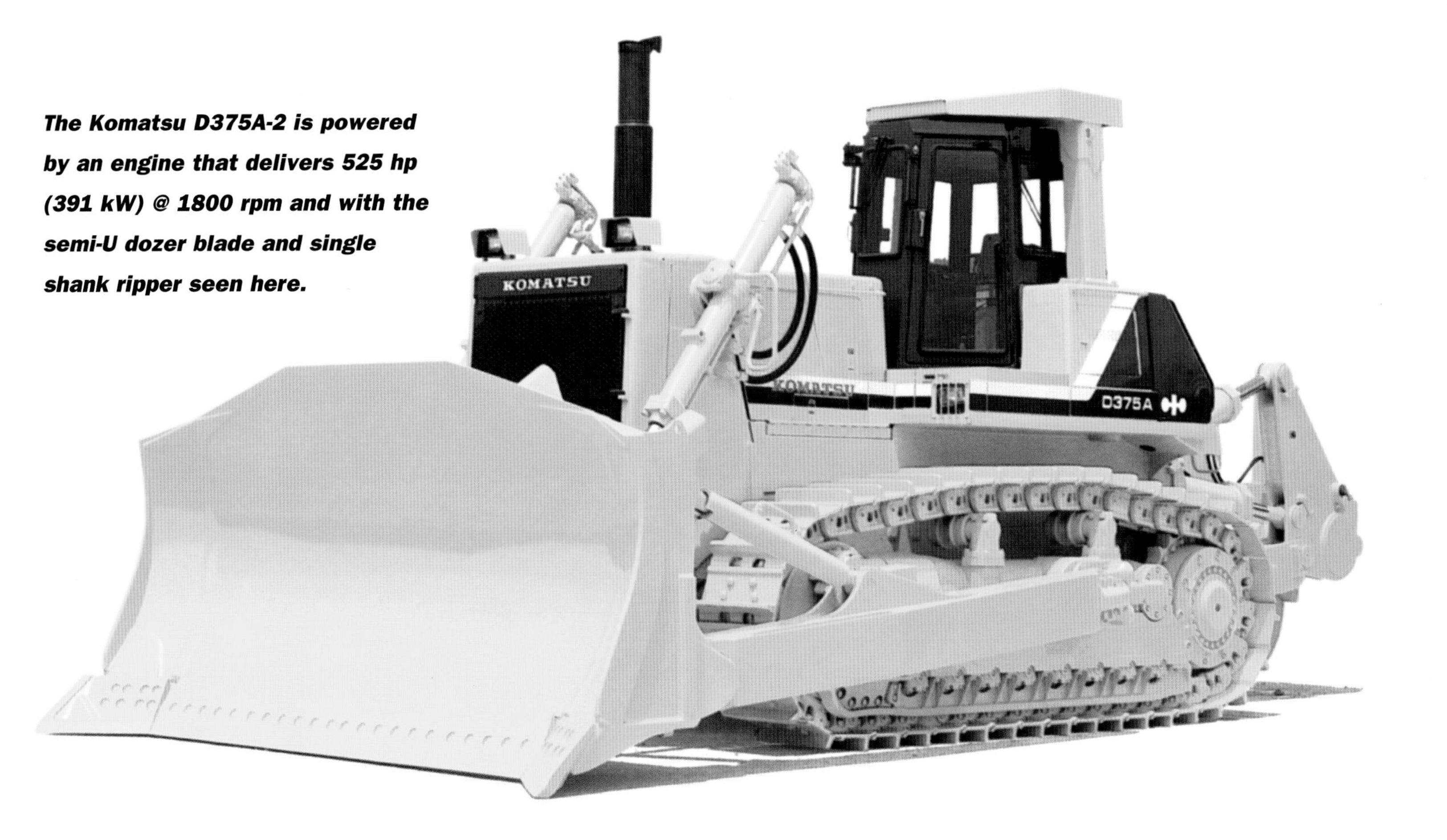


***The Komatsu D375A-2 is powered by an engine that delivers 525 hp (391 kW) @ 1800 rpm and with the semi-U dozer blade and single shank ripper seen here.***

***Liebherr use a turbocharged Cummins diesel engine to power their PR751 model, seen here with a semi-U blade and ripper attachments.***

# Liebherr Dozers

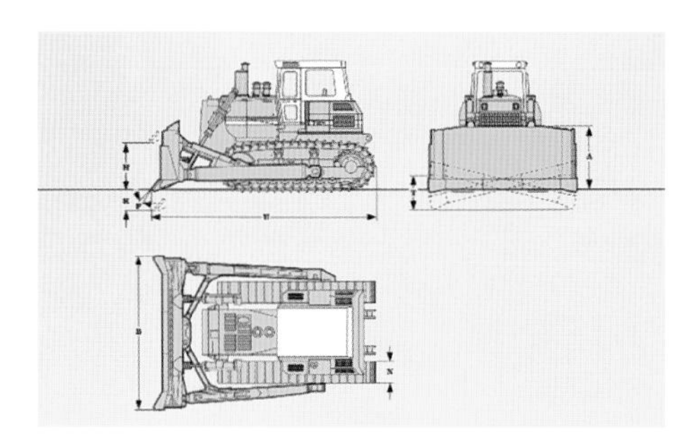

***Front and side elevations and plan of the Liebherr PR751 dozer with semi-U blade.***

**Liebherr make a range of five tracked dozers, variously designated from PR712 to PR751, the latter machine is the largest; and four crawler loaders. All are equipped with hydrostatic transmission, meaning that the dozers can be maneuvered in a non-slip manner without declutching. The drive functions are controlled by a single lever and both these features contribute to operating efficiency. The range of dozers are designed for minimum maintenance and low fuel consumption, in order to minimize operating costs.**

### Liebherr PR751 Crawler Tractor

The basic crawler tractor can be equipped with a variety of attachments—semi-U blade, U blade, angle dozer blade at the front, and either a radial single-shank ripper or multi-shank ripper at the rear. One major difference between the Liebherr PR751 and many other dozers is the fact that the Liebherr machine has a closed-loop hydrostatic travel drive which means that, rather than three specific forward and reverse gears, it has infinitely variable speeds in both forward and reverse directions. The steering system is also hydrostatic.

| LIEBHERR PR751 CRAWLER TRACTOR | |
|---|---|
| **Engine** | |
| Cummins KT 19-C | |
| **Engine power** | |
| 243 kW (330 hp) @ 1800 rpm | |
| **Bore and stroke** | |
| n/a | |
| **Transmission** | |
| Hydrostatic travel drive | |
| **Maximum speed** | |
| 5.90mph | 9.5 kph |
| **Track gauge** | |
| 69.74in | 1,770 mm |
| **Length of track on ground** | |
| 124.11in | 3,150 mm |
| **Length of basic tractor** | |
| 191.68in | 4,865 mm |
| **Maximum operating weight** | |
| 96,138lb | 43,600 kg |
| **Blade width** | |
| 160.75in | 4,080 mm |

# Other Tracked Bulldozer Manufacturers

Dozer manufacturers other than those profiled above include Case, Fiat-Hitachi, and Samsung. Fiat-Hitachi manufacture the FD145 and FD175 models, with operating weights of 15.43 and 18.74 tons (14 and 17 tonnes) respectively.

Samsung make a bigger range that includes six dozers. These range from the 6.50-ton (5.9-tonne) DX6PL to the 34.06-ton (30.9-tonne) DX30. They are made only for Korea's domestic market.

*The Samsung range of six bulldozers that vary in terms of blade width and dozing capability through the use of different capacities of engine and overall size of the machines.*

# Dozer Life

In the United States it is estimated that there are in excess of 115,000 operational dozers at work. Approximately two-thirds of this figure are owned by building, highway, and heavy construction firms, while the remainder are owned by mining and materials companies. It is believed that these companies expect 12,000 hours of use out of dozers of up to 100 horsepower and 16,500 hours from those of between 100 and 200 horsepower, but dozers of more than 200 horsepower are expected to last 20,000 hours.

*A Caterpillar dozer with the elevated drive sprocket in operation.*

***The Cat 834B wheeled dozer articulates in the center to steer like a wheeled loader. This machine is fitted with an S-blade.***

# Wheeled Dozers

***The 834B is fitted with oil cooled multiplate disc brakes which are designed as sealed adjustment-free units.***

**There is a variant of the tracked bulldozer and that is the wheeled dozer. The design of the wheeled tractor is similar to the wheeled loader because it too articulates in the center in order to steer. The center steering gives a wheeled dozer an advantage in confined places because of its maneuverability. Wheeled dozers are not manufactured by as many companies as tracked ones but are made by Caterpillar and Tiger. Caterpillar manufacture three, of which the 834B is the largest.**

**Caterpillar 834B Wheel-type Tractor**

The 834B has a minimum turning radius (over the tires) of 20.08 feet (6.12 meters) and steering is controlled by two hydraulic cylinders. Brakes are four wheeled, fully hydraulic, and enclosed wet discs. Like its crawler counterparts the wheeled dozer has both straight and U dozer blades available. Capacities vary; the straight dozer has a blade capacity of 9.51 cubic yards (7.27 cubic meters) and the U dozer 13.73 cubic yards (10.50 cubic meters).

| CATERPILLAR 834B WHEEL-TYPE TRACTOR | |
|---|---|
| **Engine** | |
| Caterpillar 3408 | |
| **Engine power** | |
| 336 kW (450 hp) @ 2100 rpm | |
| **Bore and stroke** | |
| 5.40 x 5.99in | 137 x 152 mm |
| **Transmission** | |
| 4F, 4R | |
| **Maximum speed** | |
| 13.17mph | 21.2 kph |
| **Tread** | |
| 102.09in | 2,591 mm |
| **Wheelbase** | |
| 150.11in | 3,810 mm |
| **Tire size** | |
| 35/65-33 24 PR L-4 (std) | |
| **Length of basic tractor** | |
| 304.09in | 7,718 mm |
| **Maximum operating weight** | |
| 102,168.68lb | 46,335 kg |
| **Blade width** | |
| 182.74in | 4,638 mm |

*The Tiger 690D wheeled dozer is manufactured in Australia and available with coal and semi-U (shown here) dozer blades.*

| TIGER 690D WHEEL DOZER | |
|---|---|
| **Engine** | |
| Caterpillar 3412 | |
| **Engine power** | |
| 548 kW (736 hp) @ 2200 rpm | |
| **Bore and stroke** | |
| n/a | |
| **Transmission** | |
| 3F, 3R | |
| **Maximum speed** | |
| 13.05mph | 21 kph |
| **Tread** | |
| 126.16in | 3,202 mm |
| **Tire size** | |
| 45/65 R45 XRDI type A (L4) | |
| **Wheelbase** | |
| 190.14in | 4,826 mm |
| **Length of basic tractor** | |
| 379.30in | 9,627 mm |
| **Maximum operating weight** | |
| 198,350.78lb | 89,955 kg |
| **Blade width** | |
| 244.28in | 6,200 mm |

### Tiger 690 Conversions

In 1981 Tiger Engineering Pty. Ltd. of Western Australia released what they called the Tiger 690. It was a conversion of a Caterpillar 992 wheeled loader into a wheeled dozer. The conversion was carried out by installing a new front frame designed for wheeled dozer applications, including the solid one-piece push beam, articulation plates, front axle supports, and mudguards. A number of Caterpillar components were also added including blade lift cylinders, tilt and tip cylinders, Caterpillar D10 hydraulic system, an off-highway truck torque converter, and dozer blades. The kit could be supplied for on-site assembly.

### Tiger 590B

The first Tiger 590B was released in the spring of 1994. It was based on the design of the already extant Tiger 690D but used much of the new Caterpillar 990 wheeled loader. The machine is designed for dozing large amounts of material and has proven popular with power-generating utilities where large quantities of coal needs handling—a specific coal dozer blade is available. The 590B uses the Caterpillar D9N hydraulic system, including the dual tilt and tip system.

*This Tiger 590B is fitted with a large capacity coal blade*

| TIGER 590B WHEEL DOZER | |
|---|---|
| **Engine** | |
| Caterpillar | |
| **Engine power** | |
| 503 kW (675 hp) @ 2000 rpm | |
| **Bore and stroke** | |
| n/a | |
| **Transmission** | |
| 3F, 3R | |
| **Maximum speed** | |
| 13.98mph | 22.5 kph |
| **Tread** | |
| 120.17in | 3,050 mm |
| **Tire size** | |
| 45/65 R39 XLD D1A type A (L4) | |
| **Wheelbase** | |
| 181.24in | 4,600 mm |
| **Length of basic tractor** | |
| 368.67in | 9,357 mm |
| **Maximum operating weight** | |
| 152,645.54lb | 69,227 kg |
| **Blade width** | |
| 207.95in | 5,278 mm |

TIM BUTLER
ÅKERMAN
E61

# Tracked Excavators

# Tracked Excavators

***Tracked excavators are used for highway construction in both backhoe (shown here) and face shovel configurations.***

**The boom in the development of excavators did not occur until the years after the Second World War. Alongside those experimenting with the development of the hydraulic bulldozer and loader, others were applying hydraulics to actual excavating machinery. Georges Bataille, a Frenchman, used American techniques developed during the war and introduced his first all-hydraulic design —the TU—in 1950. It was a small machine mounted on a trailer and powered by the power take-off (PTO) of the tractor towing it. Bataille displayed his machine, named Poclain after a nearby town, at the Agricultural Machine Exhibition in Paris in 1951. Other early hydraulic excavator exponents include Brödr.Söyland A/S of Norway, Blaw-Knox Ltd. of England, and NCK-Rapier Ltd. who built the Koehring 505 Skooper. Later companies include the Link-Belt Speeder Division of the American FMC Corporation, Hy-Mac in Great Britain, and Hanomag in Germany.**

***The tracked excavator is widely used in quarrying applications as seen with this Hitachi backhoe.***

Once workable designs appeared they became enormously popular, and larger and larger machines were developed with payloads that had been thought of as solely the province of cable-operated excavators. By 1971 Poclain had introduced the 169.12-ton (153.43-tonne) EC1000, which was upgraded to the 200.48-ton (181.88-tonne) 1000CK Series I in 1975. German companies, including O&K, Liebherr, and Mannesmann Demag, also produced massive, hydraulic, crawler-mounted mining excavators. O&K produced the 168-ton (152.41-tonne) RH75 in 1975 and the RH300 in 1979. Liebherr produced the 203.84-ton (184.92-tonne) R991 in 1977 and Mannesman-Demag built the H241 in 1978. In the United States Koehring, then Harnschfeger, Bucyrus-Erie, and finally Marion manufactured massive mining excavators. So successful is the use of hydraulic systems in this application that there are now countless hydraulic excavators in a variety of sizes built by manufacturers around the world, specially Europe, the United States, and Southeast Asia.

There are two distinct types of excavator: face or front shovels and backhoes. The major difference between the two is the direction in which the leading edge of the bucket faces. A backhoe draws its bucket back toward itself in order to fill it, while a front shovel pushes it away. The choice of which type of machine is most suitable depends on the type of excavation work to be carried out. The major excavator manufacturers, other than those detailed in this chapter, are Atlas, Case, Fiat-Hitachi, Hyundai, and Mannesmann Demag.

# Åkerman Excavators

*Tracked excavators are useful in construction work because of their ability to cross difficult and muddy terrain.*

**Åkerman is one of the companies within the Volvo Construction Equipment Group. Their range of seven crawler excavators is built to be durable in difficult operating environments. The machines feature three-circuit hydraulics, one circuit for each digging movement, and water-cooled, turbocharged diesel engines provide the motive power.**

The machines are variously designated EC130 to EC650. The "E" signifies excavator, the "C" crawler, and the numerical suffix refers to the machine's operating weight. All are manufactured in a backhoe configuration. The EC200 has a choice of arms, booms, and buckets available (figures quoted in the specification sheets for this and other models in this chapter are for the smallest bucket and shortest boom). The boom is the part of the excavating equipment connected to the crawler while the arm is the section between the boom and the bucket.

The booms fitted to the EC200 are in either 15.26 or 17.06-foot (4.65 or 5.2-meter) lengths. The leverage exerted increases as the boom and arm are made longer, but the shorter boom, while having less reach, can lift a greater load. Arms are available in 7.87 and 9.19-foot (2.4 and 2.8-meter) lengths. The buckets vary for specialist tasks including shifting rock, heavy-duty purposes, bulk excavation, trenching, and ditch clearing.

*The Åkerman EC200 tracked excavator is powered by a 145 hp (107 kW) engine manufactured by Volvo.*

**ÅKERMAN EC200 CRAWLER EXCAVATOR**

| | | |
|---|---|---|
| **Engine** | Volvo TD 61GE | |
| **Engine power** | 107 kW (145 hp) @ 1800 rpm | |
| **Bore and stroke** | 3.88 x 4.73in | 98.43 x 120 mm |
| **Transmission** | Hydraulic | |
| **Maximum speed** | 3.23mph | 5.2 kph |
| **Track gauge** | 90.62in | 2,300 mm |
| **Length of track on ground** | 136.32in | 3,460 mm |
| **Length** | 350.66in | 8,900 mm |
| **Maximum operating weight** | 43,659lb | 19,800 kg |
| **Bucket cutting width** | 44.33in | 1,125 mm |
| **Maximum reach** | 27.89ft | 8.5 m |

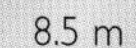

***The engine and hydraulic systems of excavators such as the Åkerman EC200 are accessed through hatches.***

**ÅKERMAN EC300 CRAWLER EXCAVATOR**

| | | |
|---|---|---|
| **Engine** | Volvo TD 71GE | |
| **Engine power** | 154 kW (209 hp) @ 2000 rpm | |
| **Bore and stroke** | 4.13 x 5.12in | 104.77 x 130 mm |
| **Transmission** | Hydraulic | |
| **Maximum speed** | 3.23mph | 5.2 kph |
| **Track gauge** | 96.53in | 2,450 mm |
| **Length of track on ground** | 165.87in | 4,210 mm |
| **Length** | 405.82in | 10,300 mm |
| **Maximum operating weight** | 68,355lb | 31,000 kg |
| **Bucket cutting width** | 47.28in | 1,200 mm |
| **Maximum reach** | 34.12ft | 10.4 m |

***The EC300 is also powered by a Volvo diesel engine of a direct injection turbocharged design.***

**Åkerman EC650 Crawler Excavator**

The EC650 is Åkerman's largest crawler excavator and is designed to be as versatile as possible. Its hydraulic system has three working modes, each of which is denoted by an acronym. ECO (economy) is the economy position for work requiring precision. CAP (capacity) is the setting for maximum capacity and speed. In this position all four working pumps supply maximum flow to the digging equipment, while the fifth pump is prioritized for swing movements. HLD (heavy lift device) is for heavy-duty digging, and in this mode the pressure to the system is increased by a corresponding degree.

The whole machine is powered by a 732-cubic inch (12-liter) capacity diesel engine that is intercooled. It is designed for maximum combustion for low emissions and maximum fuel economy. The undercarriage of the machine is designed for stability, with a width of 4.54 yards (4.15 meters)

| ÅKERMAN EC450 CRAWLER EXCAVATOR | |
|---|---|
| **Engine** | |
| Volvo TD 122 KKE | |
| **Engine power** | |
| 227 kW (304 hp) @ 1700 rpm | |
| **Bore and stroke** | |
| 5.13 x 5.91in | 130.17 x 150 mm |
| **Transmission** | |
| Hydraulic | |
| **Maximum speed** | |
| 2.30mph | 3.7 kph |
| **Track gauge** | |
| 114.26in | 2,900 mm |
| **Length of track on ground** | |
| 167.84in | 4,260 mm |
| **Length** | |
| 470.83in | 11,950 mm |
| **Maximum operating weight** | |
| 101,430lb | 46,000 kg |
| **Bucket cutting width** | |
| 63.04in | 1,600 mm |
| **Maximum reach** | |
| 37.08ft | 11.3 m |

# ÅKERMAN

***The EC450 is a third Volvo diesel engine tracked excavator with a backhoe. The machine's designation refers approximately to the machine's metric tonne operating weight.***

***The Åckerman EC650 shown here loading an articulated dump truck is the company's largest tracked excavator.***

and length of 6.56 yards (6.00 meters). The counterweight located on the rear of the excavator to counterbalance the load in the bucket is as compact as possible to allow effective operation in confined spaces. The undercarriage has built-in planetary gears and an oil bath swing ring in which the excavator turns.

The EC650 is available in two variants—the EC650 and EC650ME. The differences between the two are in the type of dipper arm and bucket fitted. The EC650 has a 2-ton/cubic yard (2-tonne/cubic meter) bucket on a 10.66-foot (3.25-meter) dipper arm giving a reach of 43.64 feet (13.3 meters), while the EC650ME has a 1.8-ton/cubic yard (1.8-tonne/cubic meter) bucket on a 9.02-foot (2.75 meter) dipper arm with a reach of 38.39 feet (11.7 meters). The purpose of the second variant is that it permits faster work cycles.

**ÅKERMAN EC650 CRAWLER EXCAVATOR**

| | |
|---|---|
| **Engine** | |
| Volvo TD 61GE | |
| **Engine power** | |
| 107 kW (145 hp) @ 1800 rpm | |
| **Bore and stroke** | |
| 3.88 x 4.73in | 98.43 x 120 mm |
| **Transmission** | |
| Hydraulic | |
| **Maximum speed** | |
| 3.23mph | 5.2 kph |
| **Track gauge** | |
| 90.62in | 2,300 mm |
| **Length of track on ground** | |
| 136.32in | 3,460 mm |
| **Length** | |
| 350.66in | 8,900 mm |
| **Maximum operating weight** | |
| 43,659lb | 19,800 kg |
| **Bucket cutting width** | |
| 44.33in | 1,125 mm |
| **Maximum reach** | |
| 27.89ft | 8.5 m |

# Caterpillar Excavators

**Caterpillar makes a massive range of more than 40 crawler mounted excavators. These vary in engine type and size from the 55 hp Caterpillar 3054 engine in their smallest excavator, the Model 307 to the 428 hp 3406C engine in the 375 L ME machine. These machines have operating weights of 177,470.3 pounds (7,650 kilograms) and 16,865.19 pounds (80,500 kilograms) respectively. Numerous machines in the range are designed and manufactured for specific tasks such as long reach models and demolition boom, are equipped models. The 325 L LR is a long reach excavator with a reach of 60 feet (18.29 meters). It is powered by the 168 hp 3116 TA Caterpillar diesel engine. The 350 L is designed for ultra high demolition and powered by the 286 hp 3306 turbocharged aftercooled diesel engine.**

***The Cat 5130 shown here with a front shovel is also available as a backhoe.***

**Caterpillar 5130 Hydraulic Shovel/Backhoe**

The Caterpillar 5130 is available in both front shovel and backhoe configuration and is primarily matched to the Caterpillar 777C rigid dump truck but can be teamed with other trucks in the 71.65 to 110.23-ton (65 to 100-tonne) class for efficient loading and hauling in mining, quarrying, and heavy construction situations. The 5130 is one of the company's massive range of tracked excavators that numbers in excess of 40 machines. They range in operating weight from 8.38 to 346.13 tons (7.6 to 314 tonnes). The 5130 operates at 194.01 tons (176 tonnes).

The 5130 utilizes a 3508 Electronic Unit Injection diesel engine, a V8 that has components such as AESC—Automatic Engine Speed Control. This cuts the engine speed from 1750 to 1350 revolutions per minute automatically if the hydraulic controls are not actuated for four seconds. This has the benefits of reducing fuel consumption and noise. The engine cooling system is also designed to minimize fuel consumption through use of a hydraulically driven, variable-speed fan. The fan operates at a minimum speed of 400 revolutions per minute, until increasing temperature operates a solenoid which increases the cooling fan speed. The engine is mounted into the box-section structural frame of the excavator which is reinforced with castings in areas of stress. Castings are fitted in the front end of the swing frame, the counterweight mounts, the boom mounts, and final drive mounts. The

***Because of the harsh working environment of excavators, Cat have designed the undercarriage to be maintenance free.***

undercarriage is maintenance free in the manner of the Caterpillar D11. It uses sealed and greased track to eliminate costly time-consuming maintenance. The stability of the machine is enhanced through use of a wide track gauge, and the track roller frames hold the undercarriage components rigidly in place. Automatic track tensioning is achieved through use of a gear pump which extends a pushrod attached to the idler. Check valves maintain the adjustment when the machine is not operating. The bucket is of box-section construction with castings in stress areas.

The two types of excavator configuration have specific design requirements. The front shovel can be supplied with a general purpose bucket, the same GP bucket with cutting edge protection, a rock bucket, and a high-density bucket. The mass excavator backhoe machine can be supplied with a GP bucket with cutting edge protection, a bucket designed for rock and penetration, a high-density bucket, and a coal or other light material bucket.

The operator station is luxuriously appointed for comfort and ease of operation. The operator's seat is adjustable for comfort and varying sizes of operator, the cab and nearby components are iso-mounted to eliminate noise and vibration. Sound levels, in-cab temperatures, and pressures are all controlled, and the machine is operated through levers. The cab is designed to maximize visibility of both the bucket and loading area. The operator can monitor the functions of the excavator through the Vital Information Management System (VIMS). Displays show pressures, temperatures, and levels fundamental to reliable operation of the excavator. The whole machine is of a modular design and can be broken down into eight modules for shipping.

**CATERPILLAR 5130 HYDRAULIC SHOVEL/BACKHOE**

| | | |
|---|---|---|
| **Engine** | | |
| Cat 3508 | | |
| **Engine power** | | |
| 608 kW (815 hp) @ 1750 rpm | | |
| **Bore and stroke** | | |
| | 6.70 x 7.45in | 170 x 190 mm |
| **Transmission** | | |
| Hydraulic | | |
| **Maximum speed** | | |
| | 2.05mph | 3.3 kph |
| **Track gauge** | | |
| | 185.97in | 4,720 mm |
| **Length of track on ground** | | |
| | 218.75in | 5,552 mm |
| **Length** | | |
| | 345.74in | 8,775 mm |
| **Front shovel** | | |
| Maximum operating weight | 38,367lb | 17,400 kg |
| Bucket cutting width | 141.84in | 3,600 mm |
| Maximum reach | 40.68ft | 12.4 m |
| **Mass excavator** | | |
| Maximum operating weight | 38,808lb | 17,600 kg |
| Bucket cutting width | 127.26in | 3,230 mm |
| Maximum reach | 48.89ft | 14.9 m |

***The operator cab of the 5130 is ergonomically designed to ensure operator productivity and comfort.***

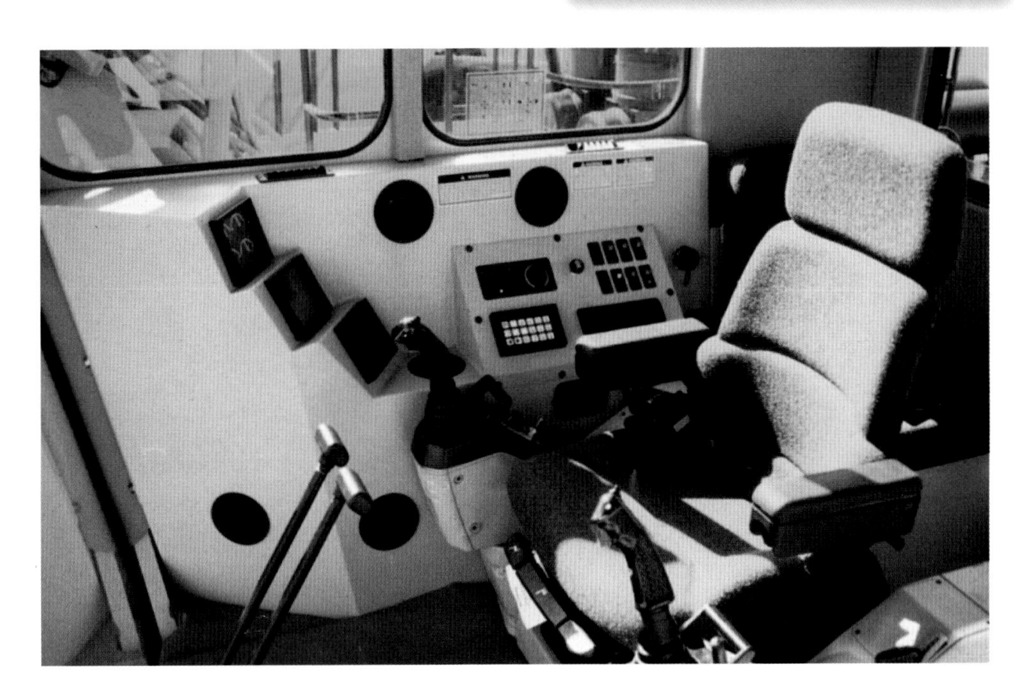

# Daewoo

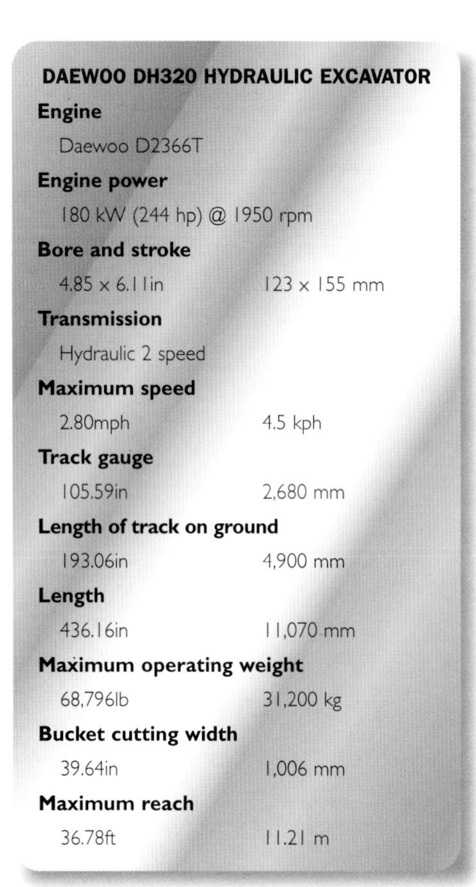

| DAEWOO DH320 HYDRAULIC EXCAVATOR | |
|---|---|
| **Engine** | |
| Daewoo D2366T | |
| **Engine power** | |
| 180 kW (244 hp) @ 1950 rpm | |
| **Bore and stroke** | |
| 4.85 x 6.11in | 123 x 155 mm |
| **Transmission** | |
| Hydraulic 2 speed | |
| **Maximum speed** | |
| 2.80mph | 4.5 kph |
| **Track gauge** | |
| 105.59in | 2,680 mm |
| **Length of track on ground** | |
| 193.06in | 4,900 mm |
| **Length** | |
| 436.16in | 11,070 mm |
| **Maximum operating weight** | |
| 68,796lb | 31,200 kg |
| **Bucket cutting width** | |
| 39.64in | 1,006 mm |
| **Maximum reach** | |
| 36.78ft | 11.21 m |

**Daewoo Heavy Industries Ltd., the Korean-based machinery company, manufacture a range of modern conventional tracked excavators. These include the DH320, Solar 400LC-III, and DH450 models which have varied operating weights ranging from 34.39 to 49.05 tons (31.2 to 44.5 tonnes). The machines are powered by Daewoo's own engines which are turbocharged, six-cylinder units of a direct-injection design.**

DAEWOO

***The Daewoo DH320 hydraulic excavator is designed for ease of operation and operator comfort.***

***Daewoo manufacture their own engines such as the D2848T (above), and a turbocharged, eight cylinder diesel for the DH450 excavator (below).***

| DAEWOO DH450 HYDRAULIC EXCAVATOR | |
|---|---|
| **Engine** | |
| Daewoo D2848T | |
| **Engine power** | |
| 221 kW (296 hp) @ 2000 rpm | |
| **Bore and stroke** | |
| 5.04 x 5.59in | 128 x 142 mm |
| **Transmission** | |
| Hydraulic 2 speed | |
| **Maximum speed** | |
| 3.42mph | 5.5 kph |
| **Track gauge** | |
| 107.96in | 2,740 mm |
| **Length of track on ground** | |
| 212.76in | 5,400 mm |
| **Length** | |
| 471.22in | 11,960 mm |
| **Maximum operating weight** | |
| 98,122.50lb | 44,500 kg |
| **Bucket cutting width** | |
| 48.86in | 1,240 mm |
| **Maximum reach** | |
| 40.03ft | 12.20 m |

***A Halla HE 220LC tracked excavator with the engine and cab hatches open.***

# Halla Engineering Tracked Excavators

**Another Korean company that manufacture tracked excavators is Halla Engineering and Heavy Industries Ltd. This company make several different machines including the HE 220LC, HE 280LC, and a relatively new machine the HE 360LCH. Halla use Cummins diesel engines to power their products, and feature three operational modes—H, S, and L, depending on the severity of excavation to be done.**

**HALLA HE 220LC HYDRAULIC EXCAVATOR**

**Engine**
Cummins 6BTA5.9-C

**Engine power**
114 kW

**Bore and stroke**
4.02 x 4.73in — 102 x 120 mm

**Transmission**
Hydraulic 2 speed

**Maximum speed**
3.54mph — 5.7 kph

**Track gauge**
94.17in — 2,390 mm

**Length of track on ground**
143.42in — 3,640 mm

**Length**
392.03in — 9,950 mm

**Maximum operating weight**
46,305lb — 21,000 kg

**Bucket cutting width**
n/a

**Maximum reach**
33.17ft — 10.11 m

***The Halla HE 280LC is the mid-range model of tracked excavator from Halla's range of three machines.***

**HALLA HE 280LC HYDRAULIC EXCAVATOR**

**Engine**
Cummins 6BTA5.9-C

**Engine power**
134 kW

**Bore and stroke**
4.49 x 5.31in — 114 x 135 mm

**Transmission**
Hydraulic 2 speed

**Maximum speed**
2.98mph — 4.8 kph

**Track gauge**
102.44in — 2,600 mm

**Length of track on ground**
158.78in — 4,030 mm

**Length**
420.40in — 10,670 mm

**Maximum operating weight**
62,181.00lb — 28,200 kg

**Bucket cutting width**
n/a

**Maximum reach**
35.57ft — 10.84 m

***The HE 360LCH is Halla's largest tracked excavator and is powered by a Cummins diesel engine.***

**HALLA HE 360LCH HYDRAULIC EXCAVATOR**

**Engine**
Cummins LTA 10-C

**Engine power**
190.23 kW (255 hp) @ 1800 rpm

**Bore and stroke**
n/a

**Transmission**
Hydraulic 2 speed

**Maximum speed**
2.98mph — 4.8 kph

**Track gauge**
107.96in — 2,740 mm

**Length of track on ground**
171in — 4,340 mm

**Length**
435.76in — 11,060 mm

**Maximum operating weight**
79,380lb — 36,000 kg

**Bucket cutting width**
n/a

**Maximum reach**
36.85ft — 11.23 m

*The JCB JS160LC is one of JCB's middleweight tracked excavators and like most others is reliant on three hydraulic assemblies to operate as a backhoe.*

# JCB Tracked Excavators

*The backhoe is operated by means of levers conveniently placed at the ends of the operator's armrests while the motion functions of the machine are controlled by the pairs of joysticks and foot pedals.*

**The famous British plant manufacturer JCB produce a range of eight different tracked excavators. These range from the JS70, a compact machine designed specially for operation in confined spaces with an operating weight of 7.83 tons (7.1 tonnes), to the JS450LC, with an operating weight of 48.83 tons (44.3 tonnes). Amongst those in between there is the JS200, which is produced in two versions, the JS200/220, and the JS220LC Long Reach. The latter version has an extremely long boom and dipper arm to make it suitable for specialist applications such as waterway maintenance. To enable it to carry out such tasks, specialist buckets are available including one for weed mowing.**

| JCB JS160 CRAWLER EXCAVATOR | |
|---|---|
| **Engine** | |
| Isuzu A4BG1T-S2 | |
| **Engine power** | |
| 72 kW (96 hp) @ 2300 rpm | |
| **Bore and stroke** | |
| n/a | |
| **Transmission** | |
| Hydrostatic 3 speed | |
| **Maximum speed** | |
| 3.42mph | 5.5 kph |
| **Track gauge** | |
| 78.41in | 1,900 mm |
| **Length of track on ground** | |
| 121.75in | 3,090 mm |
| **Length** | |
| 328.99in | 8,350 mm |
| **Maximum operating weight** | |
| 37,264.50lb | 16,900 kg |
| **Bucket cutting width** | |
| 23.64in | 600 mm |
| **Maximum reach** | |
| 27.89ft | 8.5 m |

***A specialist application is the 49 foot (15 meter) long reach boom used for tasks such as waterway maintenance seen here on this J220LC machine.***

**JCB JS220LC LONG REACH CRAWLER EXCAVATOR**

| | | |
|---|---|---|
| **Engine** | Isuzu 6BG1T | |
| **Engine power** | 95.5 kW (128 hp) @ 2020 rpm | |
| **Bore and stroke** | n/a | |
| **Transmission** | Hydrostatic 3 speed | |
| **Maximum speed** | 3.42mph | 5.5 kph |
| **Track gauge** | 94.17in | 2,390 mm |
| **Length of track on ground** | 144.20in | 3,660 mm |
| **Length** | 493.29in | 12,520 mm |
| **Maximum operating weight** | 48,734.91lb | 22,102 kg |
| **Bucket cutting width** | 23.64in | 600 mm |
| **Maximum reach** | 51.18ft | 15.6 m |

### JCB JS450LC Crawler Excavator

This is the largest JCB excavator designed to be capable of working in the toughest environments. Motive power comes from a six-cylinder, turbocharged, diesel engine manufactured by Isuzu. It is of the direct-injection type which provides for high combustion efficiency. The engine functions are monitored by what the manufacturer terms CAPS (Computer Aided Power Control System) which continuously monitors operating functions and allows three modes to be selected; H for High Production, S for general excavation and trenching, and L for precise or lighter work. Four boom lengths and a variety of buckets are available.

**JCB JS450LC CRAWLER EXCAVATOR**

| | | |
|---|---|---|
| **Engine** | Isuzu 6RB1T PE-01 | |
| **Engine power** | 201 kW (270 hp) @ 2000 rpm | |
| **Bore and stroke** | n/a | |
| **Transmission** | Hydrostatic 2 speed | |
| **Maximum speed** | 3.11mph | 5 kph |
| **Track gauge** | 107.96in | 2,740 mm |
| **Length of track on ground** | 167.84in | 4,260 mm |
| **Length** | 476.74in | 12,100 mm |
| **Maximum operating weight** | 97,769.70lb | 44,340 kg |
| **Bucket cutting width** | 49.25in | 1,250 mm |
| **Maximum reach** | 37.17ft | 11.33 m |

***The JCB JS450LC is powered by an in-line six cylinder water cooled, turbo-charged, direct injection diesel engine.***

# Kobelco Tracked Excavators

**The excavators branded as Kobelco machines are produced by the Engineering & Machinery Division of Kobe Steel Ltd. of Tokyo. In this family of excavators are machines ranging from the 7.72-ton (7-tonne) operating weight SK60 to the 50.71-ton (46-tonne) SK460LC. Modern electronics control the majority of the functions of the machines. Kobelco call it KPSS—Kobelco Power Sensing System.**

**KOBELCO SK330LC MARK IV HYDRAULIC EXCAVATOR**

**Engine**
Cummins 6CTA8.3-240

**Engine power**
165 kW (221 hp) @ 2200 rpm

**Bore and stroke**
4.49 x 5.32in    114 x 135 mm

**Transmission**
Hydrostatic 2 speed

**Maximum speed**
3.42mph    5.5 kph

**Track gauge**
102.44in    2,600 mm

**Length of track on ground**
183.01in    4,645 mm

**Length**
431.04in    10,940 mm

**Maximum operating weight**
67,914lb    30,800 kg

**Bucket cutting width**
43.73in    1,110 mm

**Maximum reach**
34.78ft    10.60 m

***The Kobelco SK330LC tracked excavator has a bucket capacity width of 43.73 inches (111 centimeters).***

***A Kobelco tracked excavator in use for highway construction.***

KOBELCO

| KOBELCO SK460LC MARK IV HYDRAULIC EXCAVATOR | |
|---|---|
| **Engine** | |
| Cummins M11-C320 | |
| **Engine power** | |
| 228 kW (306 hp) @ 2200 rpm | |
| **Bore and stroke** | |
| 4.93 x 5.79in | 125 x 147 mm |
| **Transmission** | |
| Hydrostatic 2 speed | |
| **Maximum speed** | |
| 3.42mph | 5.5 kph |
| **Track gauge** | |
| 108.35in | 2,750 mm |
| **Length of track on ground** | |
| 201.53in | 5,115 mm |
| **Length** | |
| 464.13in | 11,780 mm |
| **Maximum operating weight** | |
| 94,594.50lb | 42,900 kg |
| **Bucket cutting width** | |
| 43.35in | 1,100 mm |
| **Maximum reach** | |
| 38.45ft | 11.72 m |

***The Kobelco SK460LC has a similar size bucket but longer reach than the SK330LC.***

# Komatsu Tracked Excavators

| KOMATSU PC95 CRAWLER EXCAVATOR | |
|---|---|
| **Engine** | |
| Perkins 100.4 | |
| **Engine power** | |
| 61.5 kW (82.5 hp) @ 2000 rpm | |
| **Bore and stroke** | |
| 3.94 x 5.00in | 100 x 127 mm |
| **Transmission** | |
| Hydrostatic | |
| **Maximum speed** | |
| 3.42mph | 5.5 kph |
| **Track gauge** | |
| 70.92in | 1,800 mm |
| **Length of track on ground** | |
| 92.20in | 2,340 mm |
| **Length** | |
| 241.52in | 6,130 mm |
| **Maximum operating weight** | |
| 20,087.55lb | 9,110 kg |
| **Bucket cutting width** | |
| n/a | |
| **Maximum reach** | |
| n/a | |

**Like JCB, Komatsu manufacture a range of tracked excavators, and include a compact in that range. Komatsu's compact is the PC95, a 10.03-ton (9.1-tonne) machine. At the other end of the range is the PC1600, a 176.37-ton (160-tonne) monster. The PC1600 and two of the smaller machines, the PC1000 and PC650, are available as both tracked excavators and tracked face shovels. The numerical suffix in the model designations equates to the machines' operating weight in tonnes. The mid-range Komatsu tracked excavator is the PC340-6, and smaller-capacity ones are the PC130, PC160, and PC180 models. All feature HydrauMind—Komatsu's patented hydraulic and electronic control system.**

***A specialist application is the 49 foot (15 meter) long reach boom used for tasks such as waterway maintenance seen here on this J220LC machine.***

***One of Komatsu's larger tracked excavators is the PC340-6 which is assembled in Europe by the Japanese manufacturer.***

**KOMATSU PC340-6 CRAWLER EXCAVATOR**

**Engine**
Komatsu

**Engine power**
173 kW @ 1750 rpm

**Bore and stroke**
n/a

**Transmission**
Hydraulic 3 speed

**Maximum speed**
3.42mph 5.5 kph

**Track gauge**
n/a

**Length of track on ground**
n/a

**Length**
n/a

**Maximum operating weight**
74,970lb 34,000 kg

**Bucket cutting width**
n/a

**Maximum reach**
26.90ft 8.2 m

# KOMATSU

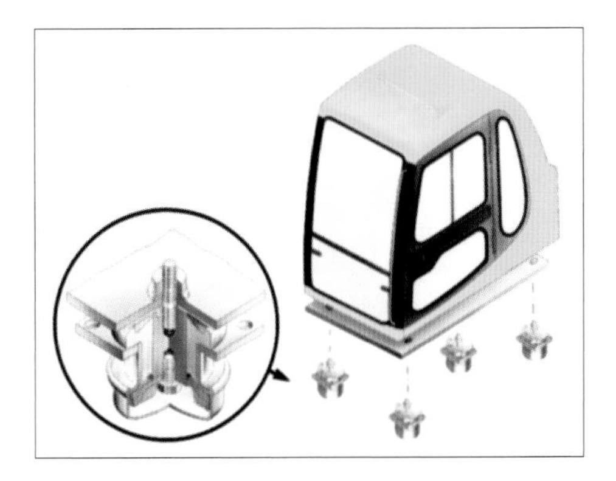

***To keep operator fatigue to a minimum, the cab rests on viscous damping mounts in order to reduce vibration and noise levels from the machine body.***

***The Liebherr tracked excavator is one of several machines at work on this busy construction site.***

# Liebherr

***Liebherr's massive mining-class face shovel, the R996 Litronic with its 36.62 cubic yard (28 cubic meter) bottom dump bucket.***

**Liebherr manufacture a range of 16 tracked excavators of which several are available as face shovels, including the R964, R974B, R984B, R992, and the R996. This latter machine is the biggest in Liebherr's range; it is a mining-class hydraulic excavator with a 584.22-ton (530-tonne) operating weight.**

**Liebherr R996 Litronic**

This excavator can be equipped with either a face shovel or backhoe attachment, with a variety of buckets for different applications and material characteristics depending upon the task. In standard form it is equipped with a heavy-duty bottom dump bucket with a 36.62-cubic yard (28.00-cubic meter) capacity suitable for rock and overburden excavation. The bucket is made from wear-resistant steel and further protected through the addition of wear-resistant plates. The bucket jaw is hydraulically operated.

The undercarriage consists of a three-piece assembly designed to be easy to transport. The side frames and car body are made of steel, and are easy to assemble and disassemble for transport. The tracks consist of combined pad-link crawler types

**LIEBHERR R996 LITRONIC HYDRAULIC EXCAVATOR**

| | |
|---|---|
| **Engine** | |
| Two Cummins K1800E | |
| **Engine power** | |
| 2240 kW (3,000 hp) @ 1800 rpm | |
| **Bore and stroke** | |
| 6.26 x 6.26in | 159 x 159 mm |
| **Transmission** | |
| Hydraulic | |
| **Maximum speed** | |
| 1.37mph | 2.2 kph |
| **Track gauge** | |
| 236.40in | 6,000 mm |
| **Length of track on ground** | |
| 277.30in | 7,038 mm |
| **Length** | |
| 482.26in | 12,240 mm |
| **Front shovel** | |
| Maximum operating weight | |
| 1,205,253lb | 546,600 kg |
| Bucket cutting width | |
| 163.51in | 4,150 mm |
| Maximum reach | |
| 51.18ft | 15.6 m |
| **Mass excavator** | |
| Maximum operating weight | |
| 1,211,647.50lb | 549,500 kg |
| Bucket cutting width | |
| 163.51in | 4,150 mm |
| Maximum reach | |
| 65.62ft | 20 m |

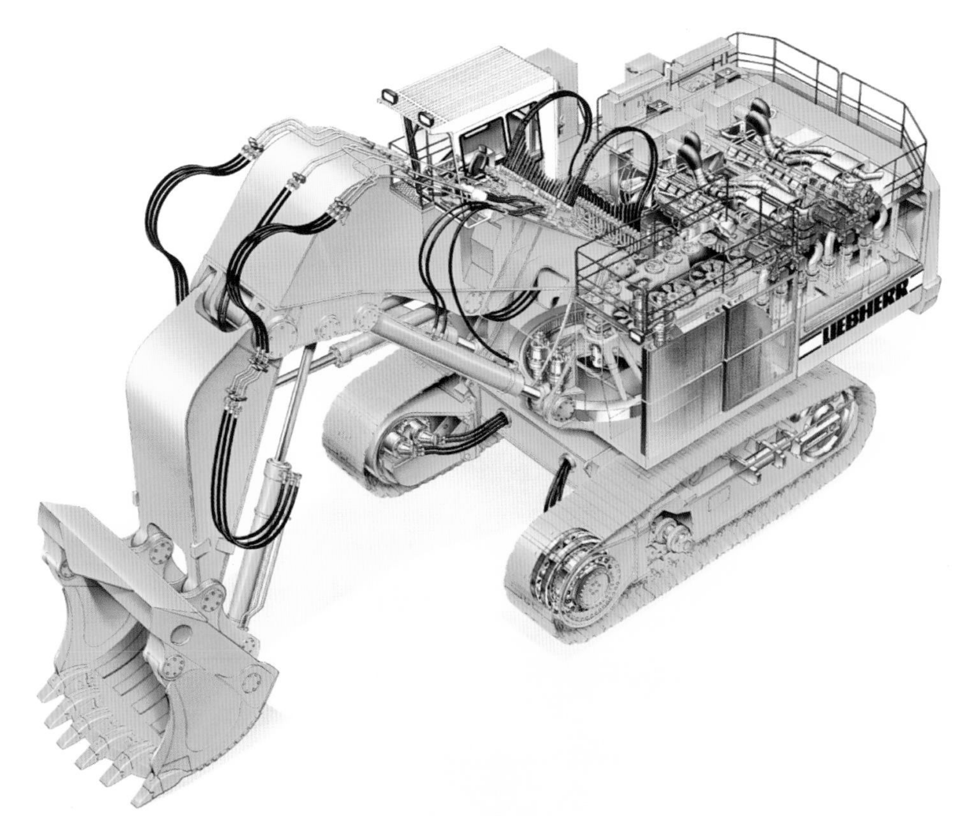


***This illustration of the R996 Litronic shows the twin engines in the superstructure, the swing ring on the undercarriage, and the hydraulic systems for both travel and shovel operation.***

assembled from maintenance-free components including cast double grouser pads. Each track has seven track rollers and three carrier rollers, and is tensioned by integral hydraulic cylinders. Independent travel drive units power the tracks and these consist of hydraulic axial piston motors and Liebherr planetary reduction gears. Braking is by wet multiplate discs. The swing ring is of an automatically lubricated, triple-roller design, and the swing-ring tower and flange is a single casting. The center of gravity of the machine is low because 40 percent of the excavator's weight is in the undercarriage.

The upper frame is a welded steel structure, made from box-section units, and carries a pair of Cummins K1800E water-cooled, turbocharged, after-cooled, direct-injection diesel engines. A 24-volt starting system and 3,435-gallon (13,002.85 liters) fuel tank are connected to the engines. The operator's cab is positioned to the front right corner of the upper superstructure for maximum visibility. It is high enough for the operator to see the loading of 264.55-ton (240-tonne) dump trucks. The engines drive a variety of hydraulic pumps including four main working ones and two swing pumps. These, and the subsidiary pumps, are controlled by Liebherr's own electronic engine speed sensing control. The R996 Litronic can load a 240.30-ton (218-tonne) truck with less than five passes.

# O&K Hydraulic Excavators

***O&K machines at work; the RH 30E tracked excavator loading a 40.5 rigid dump truck in a typical application.***

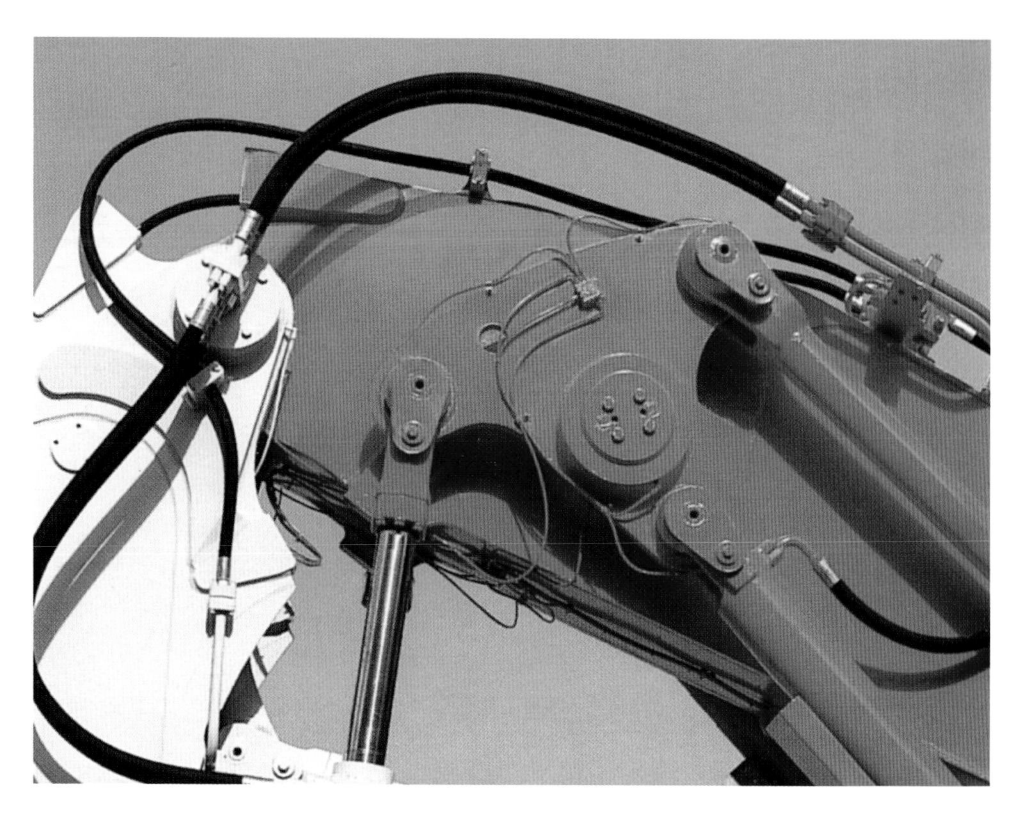

***O&K term their own design of hydraulic linkage for the boom 'TriPower Plus' referring to the three hydraulic systems required.***

| O&K RH30E HYDRAULIC EXCAVATOR | |
|---|---|
| **Engine** | |
| Cummins KT 19 C | |
| **Engine power** | |
| 336 kW (450 hp) @ 2100 rpm | |
| **Bore and stroke** | |
| n/a | |
| **Transmission** | |
| Hydraulic | |
| **Maximum speed** | |
| 1.71mph | 2.75 kph |
| **Track gauge** | |
| 137.90in | 3,500 mm |
| **Length of track on ground** | |
| 177.69in | 4,510 mm |
| **Length** | |
| 275.80in | 7,000 mm |
| **Front shovel** | |
| Maximum operating weight | |
| 173,313lb | 78,600 kg |
| Bucket cutting width | |
| 94.56in | 2,400 mm |
| Maximum reach | |
| n/a | |
| **Mass excavator** | |
| Maximum operating weight | |
| 168,682.50lb | 76,500 kg |
| Bucket cutting width | |
| 98.50in | 2,500 mm |
| Maximum reach | |
| 36.62ft | 11.16 m |

**Orenstein & Koppel AG manufacture 16 different hydraulic tracked excavators of which, in terms of operating weight, the largest five are available as either face shovels or backhoes. The RH30E is one of the biggest excavators available in the 77.16-ton (70-tonne) class. It features what O&K call "TriPower Plus"—the company's own design of hydraulic linkage for the boom, arm, and bucket operation.**

Since 1970 O&K have delivered in excess of 460 giant excavators with service weights greater than 110.23 tons (100 tonnes). There have, for example, been over 100 RH120C excavators delivered since 1984.

**O&K Hydraulic Excavators**

Like the smaller excavator models in the range, including the R90C and R120C models, the RH170 and RH200 hydraulic excavators are based around the twin-engine concept and feature the O&K TriPower attachment where bucket, crowd, and boom cylinders are connected. The machines also have a three-circuit hydraulic system, a pump management system, automatic central lubrication, and electronic control systems.

***An RH 170 hydraulic excavator loading rigid dump trucks, in this case a Cat 785.***

**O&K RH170C HYDRAULIC EXCAVATOR**

**Engine**
2 Cummins KTA38 C 925

**Engine power**
1380 kW (1,850 hp) @ 1800 rpm

**Bore and stroke**
n/a

**Transmission**
Hydraulic

**Maximum speed**
1.62mph 2.6 kph

**Track gauge**
212.76in 5,400 mm

**Length of track on ground**
244.28in 6,200 mm

**Length**
433.40in 11,000 mm

**Front shovel**
Max operating weight
749,700lb 340,000 kg
Bucket cutting width
130.02in 3,300 mm
Maximum reach
50.20ft 15.3 m

**Mass excavator**
Max operating weight
760,725lb 345,000 kg
Bucket cutting width
166.27in 4,220 mm
Maximum reach
58.93ft 17.96 m

**O&K RH90 HYDRAULIC EXCAVATOR**

**Engine**
Cummins KTA 19 C 525

**Engine power**
784 kW (1,050 hp) @ 2100 rpm

**Bore and stroke**
n/a

**Transmission**
Hydraulic

**Maximum speed**
1.71mph 2.75 kph

**Track gauge**
137.90in 3,500 mm

**Length of track on ground**
217.09in 5,510 mm

**Length**
354.60in 9,000 mm

**Front shovel**
Max operating weight
354,343.50lb 160,700 kg
Bucket cutting width
122.14in 3,100 mm
Maximum reach
41.01ft 12.5 m

**Mass excavator**
Max operating weight
359,415lb 163,000 kg
Bucket cutting width
100.47in 2,550 mm
Maximum reach
51.18ft 15.6 m

***An RH90C face shovel at work in a quarrying situation.***

# RH200 at Work

***The massive O&K RH200 tracked backhoe excavator dwarfs the rigid dump truck it is loading in an open cast coal mine.***

**O&K RH200 Hydraulic Excavators In Operation**

Currently the biggest hydraulically powered excavator in O&K's range—although an RH400 is in design—is the RH200 which was announced in 1989. Since then almost 40 units have been supplied to clients. Many of the machines have gone to clients in Australia. The Argyle Diamond Mine Pty. Ltd. of Western Australia have three RH200 face shovel machines digging blasted overburden. On average they shift 2,976.24 tons (2,700 tonnes) per hour loading trucks in four passes. The Roche Brothers Pty. Ltd. excavate gold ore at the Kalgoorlie Consolidated Gold Mines Fimiston Super Pit Gold mine in Kalgoorlie, Western Australia. The three RH200 machines are employed loading 168 and 218.40-ton (152.41 and 198.13-tonne) trucks in four and six passes respectively. Production averages out at 3,306.93 tons (3,000 tonnes) per hour. Callide Coalfields Ltd. of Queensland, Australia, have a coal mine at Biloela which is situated southwest of Gladstone, Queensland, where a single RH200 is excavating shock blasted overburden and loading 195.11-ton (177-tonne) capacity trucks. Its long-term average performance is 4,585.61 tons (4,160 tonnes) per hour. Still in Australia, the Hamersley Iron Pty. Ltd. of Perth, Western Australia, have four RH200 machines excavating blasted overburden and iron ore. Loading trucks on a five- to six-pass basis, the

average hourly production exceeds 3,858 tons (3,500 tonnes). In Canada, Syncrude Canada Ltd. have a site at Mildred Lake, Fort McMurray in Alberta, where oilsand is mined by an RH200. Production is equivalent to 3,417.16 tons (3,100 tonnes) per hour. Other O&K RH200 hydraulic excavators are in use in France, Chile, South Africa, Hong Kong, Great Britain, Venezuela, and Papua New Guinea where other applications include nickel and platinum extraction.

**O&K RH200 HYDRAULIC EXCAVATOR**

**Engine**
2 Cummins KTA38 C 1200

**Engine power**
1790 kW (2,400 hp) @ 2100 rpm

**Bore and stroke**
n/a

**Transmission**
Hydraulic

**Maximum speed**

| | |
|---|---|
| 1.43mph | 2.3 kph |

**Track gauge**

| | |
|---|---|
| 220.64in | 5,600 mm |

**Length of track on ground**

| | |
|---|---|
| 252.16in | 6,400 mm |

**Length**

| | |
|---|---|
| 472.80in | 12,000 mm |

**Front shovel**

| | |
|---|---|
| Max operating weight | |
| 1,058,400lb | 480,000 kg |
| Bucket cutting width | |
| 143.81in | 3,650 mm |
| Maximum reach | |
| 52.99ft | 16.15 m |

**Mass excavator**

| | |
|---|---|
| Max operating weight | |
| 1,065,015lb | 483,000 kg |
| Bucket cutting width | |
| 140.66in | 3,570 mm |
| Maximum reach | |
| 70.54ft | 21.5 m |

***The RH200 has two Cummins engines mounted within its superstructure to provide its motive power. It is seen here as a face shovel with a bottom dump bucket.***

***RB International plc have designed variable counterbalance hydraulic excavators such as the VC20 primarily for use in the sand and gravel extraction industry.***

# RB Variable Counterbalance Hydraulic Excavators

| RB VC20 DEEP DIG EXCAVATOR | |
|---|---|
| **Engine** | |
| n/a | |
| **Engine power** | |
| 116 kW | |
| **Bore and stroke** | |
| n/a | |
| **Transmission** | |
| Hydraulic | |
| **Maximum speed** | |
| 1.62mph | 2.6 kph |
| **Maximum operating weight** | |
| 75,631.50lb | 34,300 kg |
| **Bucket cutting width** | |
| n/a | |
| **Maximum reach** | |
| 65.62ft | 20 m |

**RB International plc build a variable counterbalance series of eight excavators that have the advantages of both a dragline and a hydraulic backhoe. The series is designed to optimize both capacity and reach through use of a moving counterweight. It is designed primarily for use in the sand and gravel industries where extraction has to take place but low ground pressure is paramount. The VC30 is the largest of the range manufactured by RB. It has a reach of up to 82.03 feet (25 meters). A deep-dig boom is available for the VC20 which gives up to 50 percent more capability than standard length booms.**

***The extremely long reach of the VC20's boom makes it suitable for coastal work such as sea defense construction.***

# Samsung Excavators

**The Korean-based Samsung Group have as one of their affiliated companies Samsung Heavy Industries who produce construction equipment, general cargo ships, power plants, and fighter planes. Samsung exports worldwide through subsidiary companies such as Samsung Construction Equipment America Corporation which is based in Illinios. This company markets a range of excavators and wheeled loaders in the United States through 150 dealers. The company has a series of six crawler excavators, ranging from the 14.33-ton (13-tonne) SE130LC-2 to the 48.17-ton (43.7-tonne) SE450LC-2. The machines feature Cummins diesel engines, with four-cylinder units in the smaller machines and six-cylinder units in the larger ones.**

Many of the excavators' functions are electronically controlled, including the swing speed, engine speed, deceleration, and mode selection system. The Samsung excavators have three operating modes: H, S, and L. These are for heavy duty, medium duty, and light duty respectively. The whole machine is designed for easy servicing. The cab is ergonomically designed and features a distinctive curved front screen which enhances the operator's visibility. The undercarriage is designed to permit two-speed travel—low gear for rough terrain and high gear for travel between job areas.

***A Korean manufactured Samsung tracked backhoe excavator in a typical working environment.***

| SAMSUNG SE210LC-2 CRAWLER EXCAVATOR | |
|---|---|
| **Engine** | |
| Cummins 6BT5.9C | |
| **Engine power** | |
| 102 kW (136 hp) @ 2100 rpm | |
| **Bore and stroke** | |
| 4.02 x 4.73in | 102 x 120 mm |
| **Transmission** | |
| Hydraulic 2 speed | |
| **Maximum speed** | |
| 2.86mph | 4.6 kph |
| **Track gauge** | |
| 94.56in | 2,400 mm |
| **Length of track on ground** | |
| 143.42in | 3,640 mm |
| **Length** | |
| 385.33in | 9,780 mm |
| **Maximum operating weight** | |
| 47,958.75lb | 21,750 kg |
| **Bucket cutting width** | |
| 43.34in | 1,100 mm |
| **Maximum reach** | |
| 32.81ft | 10 m |

| SAMSUNG SE450LC-2 CRAWLER EXCAVATOR | |
|---|---|
| **Engine** | |
| Cummins LTA 10 C | |
| **Engine power** | |
| 221 kW (296 hp) @ 2000 rpm | |
| **Bore and stroke** | |
| 4.93 x 5.36in | 125 x 136 mm |
| **Transmission** | |
| Hydraulic 2 speed | |
| **Maximum speed** | |
| 2.67mph | 4.3 kph |
| **Track gauge** | |
| 113.08in | 2,870 mm |
| **Length of track on ground** | |
| 172.18in | 4,370 mm |
| **Length** | |
| 470.83in | 11,950 mm |
| **Maximum operating weight** | |
| 96,446.70lb | 43,740 kg |
| **Bucket cutting width** | |
| 56.89in | 1,444 mm |
| **Maximum reach** | |
| 39.50ft | 12.04 m |

***The Samsung SE210LC-2 is produced in Korea at Samsung's Changwon assembly plant where Cummins diesel engines are fitted.***

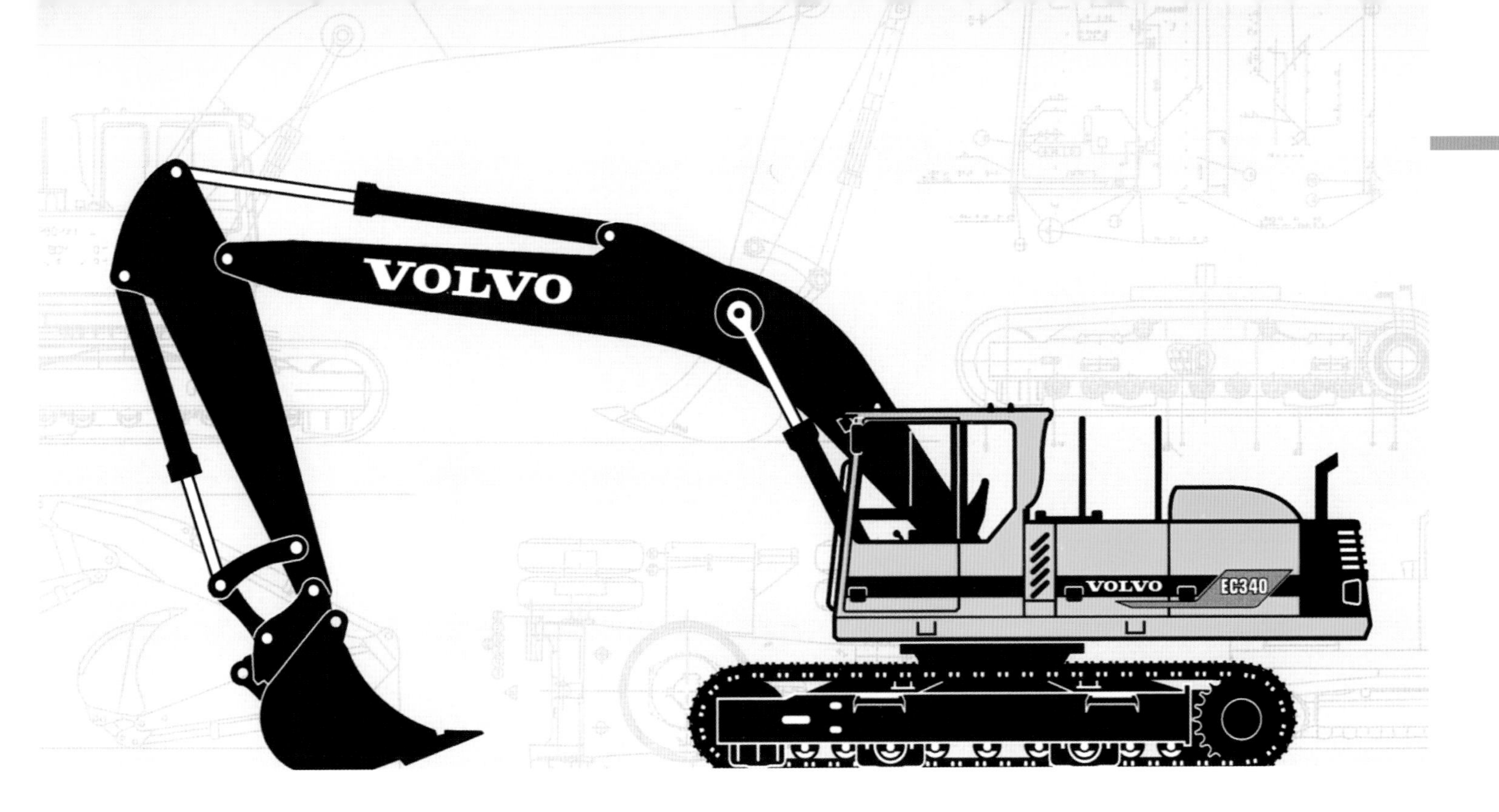


# Volvo Tracked Excavators

*Side elevation of the Volvo EC340 tracked excavator showing the three hydraulic circuits required for boom and bucket operation.*

**Volvo tracked excavators are some of the machines manufactured by the division of AB Volvo that specializes in construction equipment. Germany and the United States are the two biggest markets for Volvo construction equipment although it is sold in many other countries throughout the world.**

### Volvo EC340 Excavator

Although Åkerman is part of the Volvo Construction Equipment, Volvo produce an excavator that is not badged as an Åkerman but as a Volvo, the EC340. It is powered by a low-emission, turbocharged, direct-injection diesel engine. Much of its operation is electronically controlled, including mode selector and speed sensing control. The machine features three hydraulic circuits for boom and arm control, maneuvering precision, and fuel economy. The slew system features an axial piston motor and planetary gearbox, and the slew ring works in an oil bath. A variety of booms and dipper arms are available.

| VOLVO EC340 CRAWLER EXCAVATOR | |
|---|---|
| **Engine** | |
| Volvo TD 103 KAE | |
| **Engine power** | |
| 190 kW (258 hp) @ 1700 rpm | |
| **Bore and stroke** | |
| 4.75 x 5.52in | 120.65 x 140 mm |
| **Transmission** | |
| Hydraulic | |
| **Maximum speed** | |
| 3.11mph | 5.0 kph |
| **Track gauge** | |
| 94.56in | 2,400 mm |
| **Length of track on ground** | |
| 172.18in | 4,370 mm |
| **Length** | |
| 413.70in | 10,500 mm |
| **Maximum operating weight** | |
| 80,923.50lb | 36,700 kg |
| **Bucket cutting width** | |
| n/a | |
| **Maximum reach** | |
| 32.81ft | 10 m |

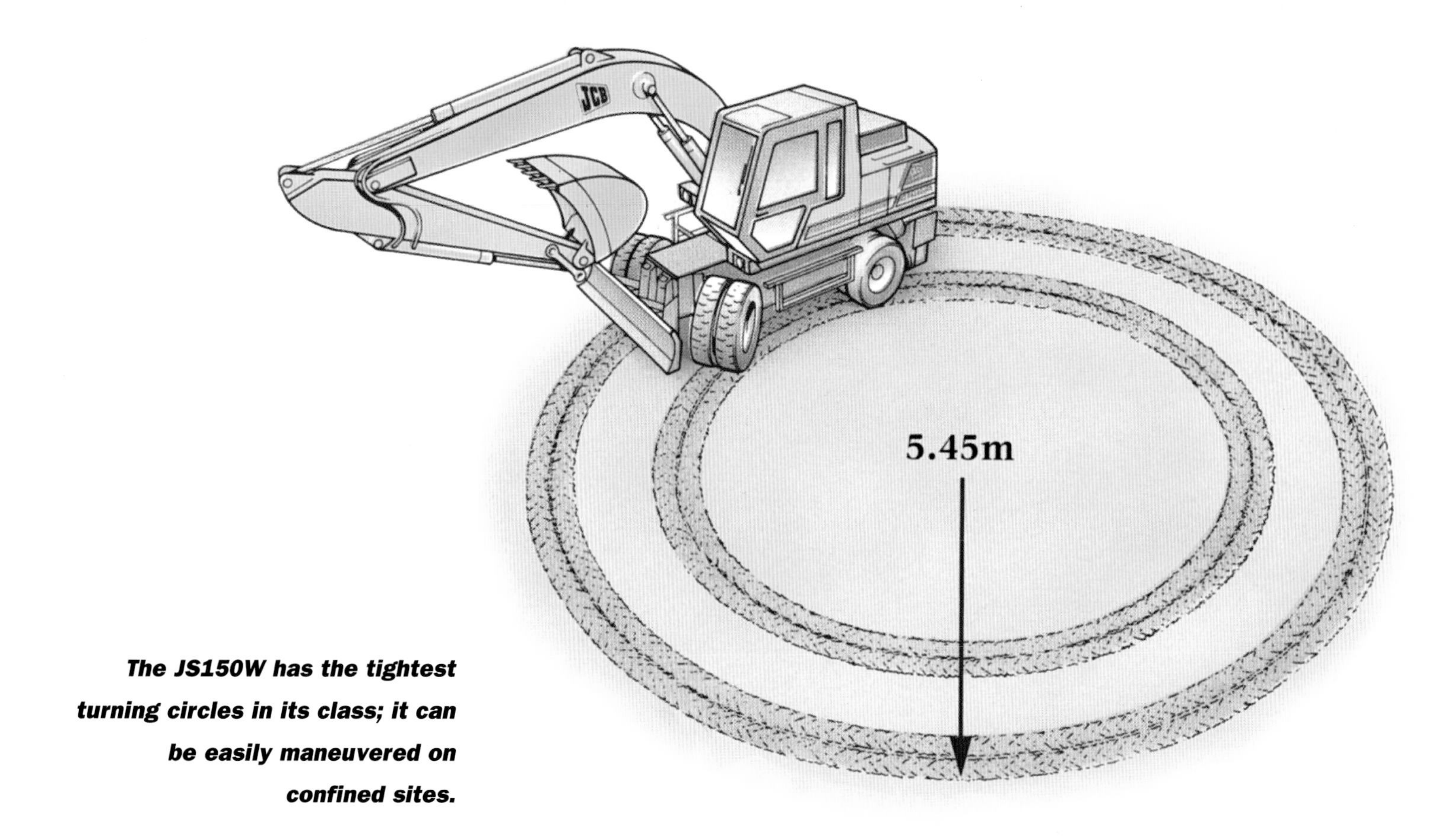


***The JS150W has the tightest turning circles in its class; it can be easily maneuvered on confined sites.***

# Wheeled Excavators

**Wheeled excavators are hydraulic excavators that carry out the same tasks as tracked ones but, in place of crawler tracks, have wheels. These are used in situations where crawlers are not considered necessary. As a rule of thumb, wheeled excavators tend to be smaller in capacity than tracked ones and rarely exceed the 22.05-ton (20-tonne) operating weight. They are manufactured by a huge range of companies including Åkerman, Atlas, Case, Caterpillar, Daewoo, Halla, Fiat-Hitachi, JCB, Komatsu, Libra, Liebherr, Mecalalc, O&K, Pel-Job, and Samsung.**

Two typical examples of the wheeled excavator are the Åkerman EW200 and the JCB JS150W. The Åkerman EW200 backhoe is simply a wheeled version of the EC200 and shares many of its components with the exception of the undercarriage. Some of its dimensions are different because of the different undercarriage, and the wheeled variant is capable of higher travel speeds.

Like the Åkerman, the JCB JS150W is equipped with a hydraulically operated dozer blade to make it more versatile on construction sites and for crowding stockpiles. It too is faster than its tracked counterpart when high range is selected on the transmission.

***A JCB JS150W excavating a trench; foundations and services require deep trenches on many construction sites.***

***Wheeled excavators such as this Akerman EW200 tend to be used in confined spaces on some construction sites and are usually smaller than the tracked excavators from the same manufacturer.***

**JCB JS150W WHEELED EXCAVATOR**

| | | |
|---|---|---|
| **Engine** | | |
| Isuzu 4BD1 PTA-15 | | |
| **Engine power** | | |
| 72 kW (96 hp) @ 2300 rpm | | |
| **Bore and stroke** | | |
| 4.02 x 4.65in | 102 x 118 mm | |
| **Transmission** | | |
| Hydrostatic 2 speed | | |
| **Maximum speed** | | |
| 15.54mph | 25 kph | |
| **Gauge** | | |
| 52.40in | 1,330 mm | |
| **Wheelbase** | | |
| 118.20in | 3,000 mm | |
| **Length** | | |
| 327.81in | 8,320 mm | |
| **Maximum operating weight** | | |
| 35,081.55lb | 15,910 kg | |
| **Bucket cutting width** | | |
| 19.70in | 500 mm | |
| **Maximum reach** | | |
| 27.89ft | 8.5 m | |

**ÅKERMAN EW200 CRAWLER EXCAVATOR**

| | |
|---|---|
| **Engine** | |
| Volvo TD 61GE | |
| **Engine power** | |
| 107 kW (145 hp) @ 1800 rpm | |
| **Bore and stroke** | |
| 3.88 x 4.73in | 98.43 x 120 mm |
| **Transmission** | |
| Hydraulic | |
| **Maximum speed** | |
| 18.64mph | 30 kph |
| **Gauge** | |
| 75.25in | 1,910 mm |
| **Wheelbase** | |
| 98.50in | 2,500 mm |
| **Length** | |
| 327.02in | 8,300 mm |
| **Maximum operating weight** | |
| 39,690lb | 18,000 kg |
| **Bucket cutting width** | |
| 44.33in | 1,125 mm |
| **Maximum reach** | |
| 27.89ft | 8.5 m |

ATT
600

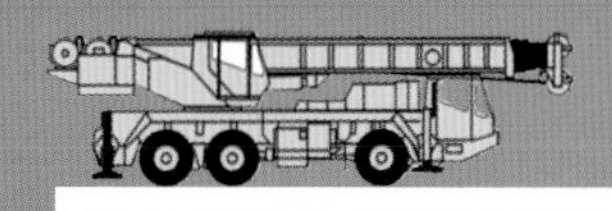

# Cranes and Draglines

# The History of Grove Cranes

***A crawler mounted crane with a lattice boom in use during highway construction.***

Grove Worldwide are manufacturers of a large and comprehensive range of cranes in the following classes.

**Rough terrain hydraulic cranes** Grove make 14 different models with rated lift capacities from 13.22 to 99.21 tons (12 to 90 tonnes).

**All terrain hydraulic cranes** Grove manufacture 16 various models with rated lift capacities ranging from 22.05 to 220.46 tons (20 to 200 tonnes).

**Truck mounted hydraulic cranes** There are 11 models in Grove's range, with lift capacities ranging from 33.07 to 165.35 tons (30 to 150 tonnes).

**Material handling cranes** Grove make six of these smaller cranes, of which the largest has a rated lift capacity of 17.53 tons (15.9 tonnes).

**Lattice boom hydraulic cranes** Grove manufacture two of these machines, one truck mounted and one crawler mounted. Both have a rated lift capacity of 149.91 tons (136 tonnes).

Grove started in business in January 1947 making rubber-tired farm wagons in a rented garage in Shady Grove, Pennsylvania. The company soon moved to a larger site, constructed a new building, and, in 1949, made their first crane. It was produced because the company needed one to assist in its own materials handling and the company was unable to buy anything suitable. The machine was refined and introduced as one of Grove Manufacturing Company's products. It was a success and other companies

***Crane booms such as this crawler mounted unit dismantle for transport between construction sites.***

ordered them. In 1959 Grove introduced their first four-wheel drive, four-wheel steer crane and a truck mounted version. In 1965 Grove sold its 1,000th crane and by June 1967 the company had made its 2,000th. In the November of the same year the company was acquired by Kidde Inc. of New Jersey and shipped its 3,000th crane in August 1968. By mid-1975 production had been increased tenfold over 1967 levels, and by January 1979 Grove had manufactured 20,000 cranes. In the same year Grove acquired Manlift Inc. of Selma, California, to add their range of boom and scissor aerial work platforms to its inventory. By 1981 Grove had produced 25,000 cranes and in February 1982 the United States government recognized Grove product quality and awarded the company the United States Department of Defense Contractor Assessment Program (CAP) Award and Flag. In September 1983 Grove launched a new product line—hydraulically powered lattice boom cranes—and received an order from Oshkosh Truck Corporation for a specially designed material handling crane to be mounted on the rear of the United States Army's HEMTT (Heavy Expanded Military Tactical Truck). Grove were to supply 6,500 cranes for this purpose. In October 1984 Coles Cranes Limited, a long-established crane manufacturer from Great Britain, was acquired by Grove. In November 1986 Grove were awarded a $50 million United States Army contract for 269 RT875CC rough terrain container cranes.

The first ever 336-ton (304.82-tonne) capacity Grove TN3000 truck crane was delivered to a Swiss operator in December 1986 and was the largest telescopic boom truck crane ever built in the United States. In November 1987 Kidde Inc. was acquired by the Hanson Trust plc from Great Britain, and a year later the AMZ66 articulating boom Grove Manlift aerial work platform was introduced. In February 1990 the company became known as Grove Worldwide, and plans for a new plant in Salem, Virginia, were announced. Their current range of all-terrain hydraulic cranes combine highway speeds of up to 55.93 miles per hour (90 kilometers per hour) with off-road performance. The rough terrain hydraulic units feature rugged box-section frames for toughness, and four-wheel steering. The truck-mounted hydraulic cranes from Grove are designed for road use and have tip heights of up to 89.68 yards (82 meters). The crawler-mounted cranes are intended for use in soft-ground applications and can be supplied with either rubber or steel tracks. Lattice boom hydraulic cranes are offered in both crawler- and carrier-mounted versions, and have tip heights of up to 120.62 yards (110.3 meters).

***A truck mounted PPM crane ready to move - note how three of the five axles steer.***

***Road construction is one task where truck mounted cranes are useful; here a Grove TM9120 lifts prefabricated concrete panels into place.***

# Truck Telescopic and Lattice Cranes

**Grove, Kato, Link-Belt, Lorain, Marchetti, Manitowoc, Mannesmann Demag, P&H, and Tadano Faun are amongst the noted manufacturers of these types of cranes. The major difference with the two types is that in the case of the telescopic cranes the boom is hydraulically actuated and telescopes out when the mechanism is actuated. The lattice type boom is fixed or hinged, depending on size, but is raised on cables. Some cranes combine both types with a telescopic boom that has a lattice jib extension.**

### Grove TM9120 Hydraulic Crane

This four-axle hydraulic crane features hydraulic outriggers for stability while lifting. The hoist is operated by a variable displacement piston motor that has pressure override for infinitely variable speed, powered in both upward and downward directions. It utilizes planetary reduction with an automatically spring applied disc brake. The drum is grooved to enable the cable to spool on easily. The boom is raised by one double-acting hydraulic cylinder with an integral holding valve. This provides the elevation from 3 degrees to 80 degrees.

| GROVE TM9120 HYDRAULIC CRANE | |
|---|---|
| **Crane type** | |
| Carrier mounted | |
| **Boom length** | |
| 42–129.93ft | 12.8–39.6 m |
| **Rated lift capacity** | |
| 132.28t | 120 MT |
| **Crane engine** | |
| Cummins 6CTA8.3 | |
| **Power** | |
| 186 kW (250 hp) @ 2200 rpm | |
| **Engine** | |
| Cummins N14-460E | |
| **Power** | |
| 343 kW (350 hp) @ 1600 rpm | |
| **Drive** | |
| 8x4 | |
| **Transmission** | |
| 10F, 3R | |
| **Maximum speed** | |
| 47.85mph | 77 kph |
| **Tire size** | |
| 14.00R24 | |

***The Kato NK-350E-V.***

***A six axle truck is used by Grove for transporting the TM1500 hydraulic crane.***

| GROVE TM1500 HYDRAULIC CRANE | |
|---|---|
| **Crane type** | |
| Carrier mounted | |
| **Boom length** | |
| 45.93–173.89ft | 14–53 m |
| **Rated lift capacity** | |
| 165.35t | 150 MT |
| **Crane engine** | |
| Cummins 6CTA8.3 | |
| **Power** | |
| 186 kW (250 hp) @ 2500 rpm | |
| **Engine** | |
| Cummins NTC 450C | |
| **Power** | |
| 335 kW (450 hp) @ 2100 rpm | |
| **Drive** | |
| 12x6 | |
| **Transmission** | |
| 9F, 2R | |
| **Maximum speed** | |
| 41.82mph | 67.3 kph |
| **Tire size** | |
| 14.00x20-22PR (G20 X ZA4) | |

**Grove TM1500 Hydraulic Crane**

The TM prefix indicates that this is a truck-mounted crane and the numerical suffix indicates that it has a rated lift capacity of 165.35 tons (150 tonnes). It is the largest of this type manufactured by Grove.

### Kato NK-350E-V Hydraulic Crane

The NK-350E-V is one of Kato's range of truck mounted telescopic hydraulic cranes that range in lifting capacity from 13.23 to 176.37 tons (12 to 160 tonnes) with the NK–120EIII and NK-1600 respectively. The NK-350E-V is in the middle of the range with its lift capacity of 38.58 tons (35 tonnes) and a boom length of 111.55 feet (34 meters). It is a four-axle machine, of which two are driven, and in traveling mode it is 513.38 inches (13,030 millimeters) long.

***Kato produce an extra long boom to increase the high lift capacity of the NK-350-E crane. The jib can be operated in three offset positions, 5 degrees, 17 degrees, and 30 degrees from vertical.***

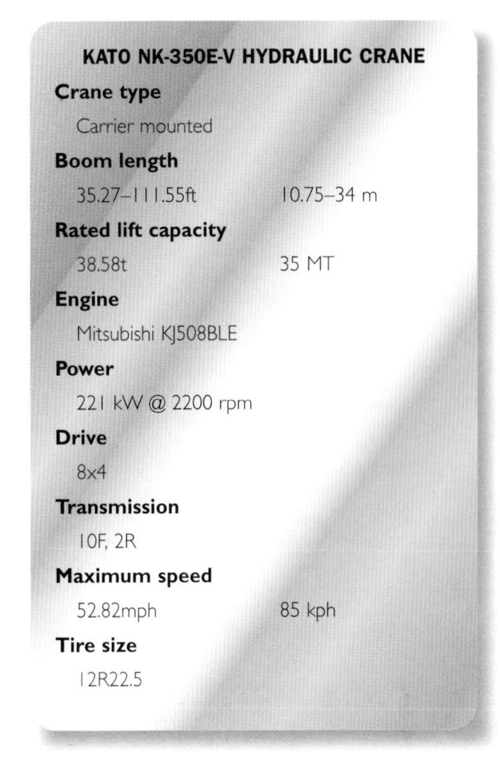

| KATO NK-350E-V HYDRAULIC CRANE | |
|---|---|
| **Crane type** | |
| Carrier mounted | |
| **Boom length** | |
| 35.27–111.55ft | 10.75–34 m |
| **Rated lift capacity** | |
| 38.58t | 35 MT |
| **Engine** | |
| Mitsubishi KJ508BLE | |
| **Power** | |
| 221 kW @ 2200 rpm | |
| **Drive** | |
| 8x4 | |
| **Transmission** | |
| 10F, 2R | |
| **Maximum speed** | |
| 52.82mph | 85 kph |
| **Tire size** | |
| 12R22.5 | |

***The three axle Kato hydraulic crane on a Mitsubishi truck chassis with purpose-built cab.***

***The four section boom of the Liebherr LTM1800 is transported between jobs on a semi-trailer with a steered trailing axle.***

| LIEBHERR LTM1800 HYDRAULIC CRANE | |
|---|---|
| **Crane type** | |
| Carrier mounted | |
| **Boom length** | |
| 63.32–196.86ft | 19.3–60 m |
| **Rated lift capacity** | |
| 881.84t | 800 MT |
| **Crane engine** | |
| Daimler Benz OM 423 A | |
| **Power** | |
| 300 kW (408 hp) @ 1200 rpm | |
| **Engine** | |
| Daimler Benz OM 444 A | |
| **Power** | |
| 390 kW (530 hp) @ 2100 rpm | |
| **Drive** | |
| 16x8 | |
| **Transmission** | |
| 10F, 2R | |
| **Maximum speed** | |
| n/a | |
| **Tire size** | |
| 14.00R24 | |

**Liebherr LTM1800 Hydraulic Crane**

This is one of the biggest mobile cranes manufactured, with its load capacity of 881.84 tons (800 tonnes), a lifting height of up to 439.65 feet (134 meters)—using both the hydraulic boom and lattice jib. It can work with either or both booms, and for travel the telescopic boom is removed and loaded onto a semi-trailer. This means the LTM1800 is then under the maximum weight limit for road travel. The LTM1800 uses its own erecting bracket to attach and detach the telescopic boom. It is assembled on an eight-axle chassis of which six axles steer.

**Mannesmann Demag AC1600 Hydraulic Crane**

The nine-axle Demag AC1600 is roadable because it meets the maximum weight of 13.23 tons (12 tonnes) per axle, with only the counterweight needing to be carried separately. Although the AC1600 has nine axles, its design is compact so that its overall length is similar to many eight-axle machines that do not have a 551.15-ton (500-tonne) lifting capacity. The crane is intended to be ready for use on site quickly; its outrigger controls are located on both sides of the carrier and it can install its own 154.32-ton (140-tonne) counterweight without an auxiliary crane. The crane cab can be tilted in order to improve the operator's view of the boom. The AC1600 is designed for attachments such as a luffing jib and Demag's patented Superlift system. A particular variant of the AC1600 has been built for Japan in order to conform to road traffic regulations in that country. The Japanese model has only eight axles and is known as the AC1600J.

Both the carrier and the crane use Daimler Benz engines; a 10-cylinder unit in the carrier and a six-cylinder in the crane, both are watercooled.

***The nine-axle Mannesmann Demag AC1600 with its hydraulic boom in the traveling position.***

**MANNESMANN DEMAG AC1600 HYDRAULIC CRANE**

| | | |
|---|---|---|
| **Crane type** | Carrier mounted | |
| **Boom length** | 51.84–164.05ft | 15.8–50 m |
| **Rated lift capacity** | 881.84t | 800 MT |
| **Crane engine** | Daimler Benz OM 443 LA | |
| **Power** | 412 kW (560 hp) | |
| **Engine** | Daimler Benz OM 447 A | |
| **Power** | 210 kW (286 hp) | |
| **Drive** | 18x8 | |
| **Transmission** | ZF Transmatic | |
| **Maximum speed** | n/a | |
| **Tire size** | 14.00R25 | |

***Mannesmann Demag AC1600 crane being used in an industrial application.***

| TADANO TG-500E | |
|---|---|
| **Crane type** | |
| Carrier mounted | |
| **Boom length** | |
| 35.93–131.24ft | 10.95–40 m |
| **Rated lift capacity** | |
| 55.12t | 50 MT |
| **Engine** | |
| Mitsubishi | |
| **Power** | |
| n/a | |
| **Drive** | |
| 8x4 | |
| **Transmission** | |
| 5F, 1R | |
| **Maximum speed** | |
| 52.82mph | 85 kph |
| **Tire size** | |
| 12R22.5 | |

## Tadano TG-500E

Tadano manufacture a range of hydraulic truck cranes, of which the one with the largest lifting capacity is the TG-500E with a 55.12-ton (50-tonne) capability. The crane is mounted on a four-axle Mitsubishi chassis that is fitted with a watercooled, direct-injection diesel engine. The boom is a five-piece hexagonal section item. The smaller counterparts include the TL-250E and TL-350E.

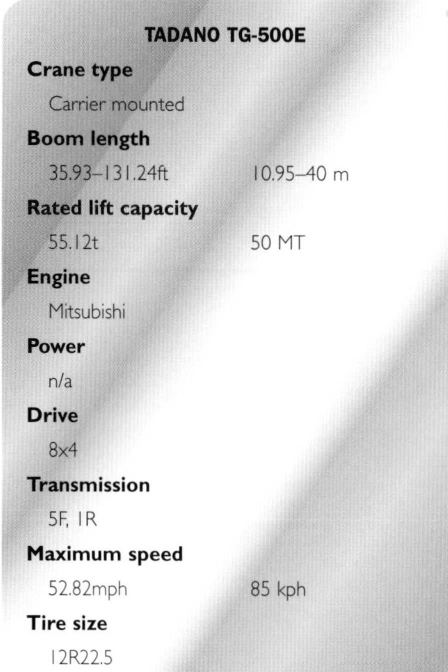

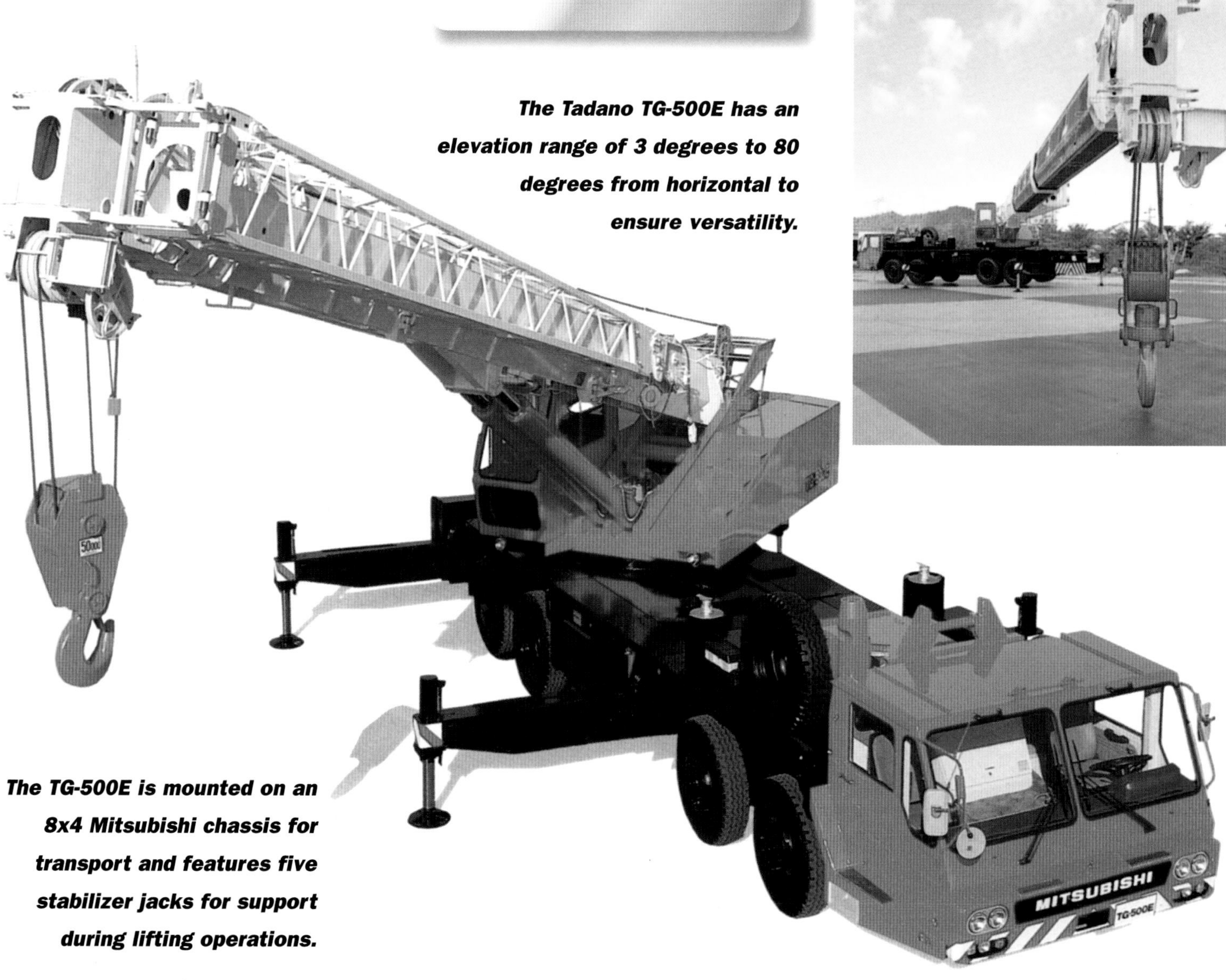


***The Tadano TG-500E has an elevation range of 3 degrees to 80 degrees from horizontal to ensure versatility.***

***The TG-500E is mounted on an 8x4 Mitsubishi chassis for transport and features five stabilizer jacks for support during lifting operations.***

***Hydraulic cranes such as this Kato have an extending boom operated by a hydraulic system.***

# All-Terrain Cranes

**Grove, Kato, Liebherr, Link-Belt, Mannesmann Demag, Marchetti, PPM, Tadano Faun, and Terex Cranes Inc. (a subsidiary of Terex Corporation) all manufacture all-terrain cranes.**

**Grove GMK 5130 Hydraulic Crane**

This five-axle machine has axles numbers two, three, and five driven, while all five steer by means of a dual-circuit, hydraulic power-assisted system. The boom is in four sections that are elevated by a single hydraulic cylinder that raises the boom from 3 degrees to 84 degrees.

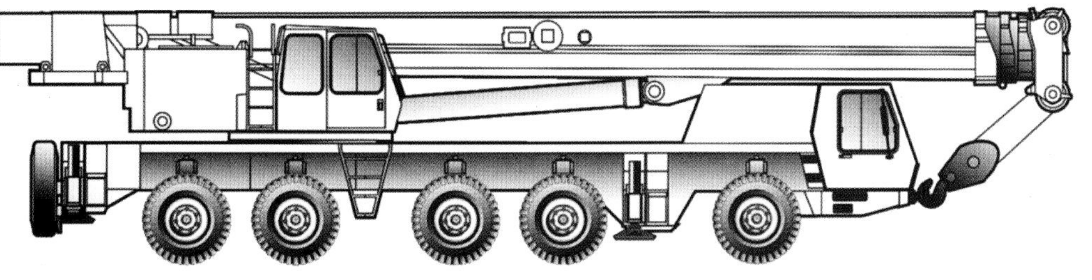

***The Grove GMK 5130 crane has five axles which all steer and three of which are driven.***

**GROVE GMK 5130 HYDRAULIC CRANE**

| | | |
|---|---|---|
| **Crane type** | | |
| | All-terrain | |
| **Boom length** | | |
| | 45.93–150.93ft | 14–46 m |
| **Rated lift capacity** | | |
| | 143.304t | 130 MT |
| **Crane engine** | | |
| | Mercedes Benz OM366A | |
| **Power** | | |
| | 104 kW @ 1800 rpm | |
| **Engine** | | |
| | Mercedes Benz OM442LA | |
| **Power** | | |
| | 370 kW @ 2100 rpm | |
| **Drive** | | |
| | 10x6 | |
| **Transmission** | | |
| | 5F, 1R | |
| **Maximum speed** | | |
| | 42.88mph | 69 kph |
| **Tire size** | | |
| | 14.00R25 | |

**Grove GMK 6200 Hydraulic Crane**

The GMK 6200 is Grove Worldwide's largest all-terrain hydraulic crane and has six axles. The carrier is powered by a 12-cylinder, turbocharged diesel engine while the crane is powered by a six-cylinder unit. Both engines are watercooled and turbocharged. The hoist is powered by an axial piston variable displacement motor with planetary gears and brake.

| GROVE GMK 6200 HYDRAULIC CRANE | |
|---|---|
| **Crane type** | |
| All-terrain | |
| **Boom length** | |
| 47.57–173.89ft | 14.5–53 m |
| **Rated lift capacity** | |
| 220.46t | 200 MT |
| **Crane engine** | |
| Mercedes Benz OM447A | |
| **Power** | |
| 213 kW @ 1800 rpm | |
| **Engine** | |
| Mercedes Benz OM444A | |
| **Power** | |
| 406 kW @ 2100 rpm | |
| **Drive** | |
| 12x8 | |
| **Transmission** | |
| 5F, 1R | |
| **Maximum speed** | |
| 41.63mph | 67 kph |
| **Tire size** | |
| 16.00R25 | |

**Marchetti MG60.3 Hydraulic Crane**

The MG60.3 is a three-axle design of all-terrain crane, with two driven axles and three steering. A turbocharged intercooler diesel from Iveco provides the power for the crane. The hydraulic equipment includes four pumps to control the operation of the crane and its outriggers. The boom is octagonal in section and in four hydraulic parts.

***Side elevation of the Grove Crane GMK 6200 all terrain hydraulic crane.***

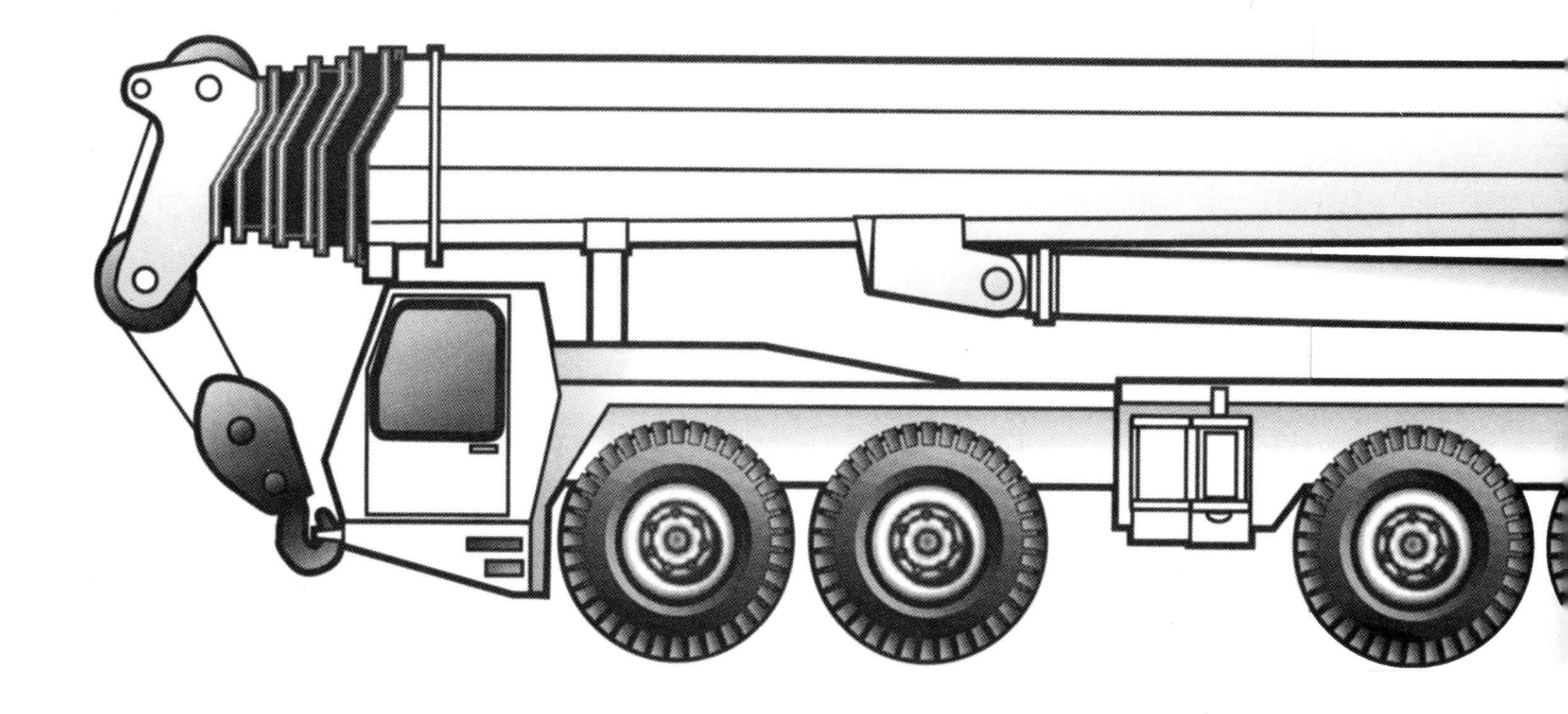

***The three axle Marchetti MG60.3 hydraulic crane prepared for travel between lifting operations.***

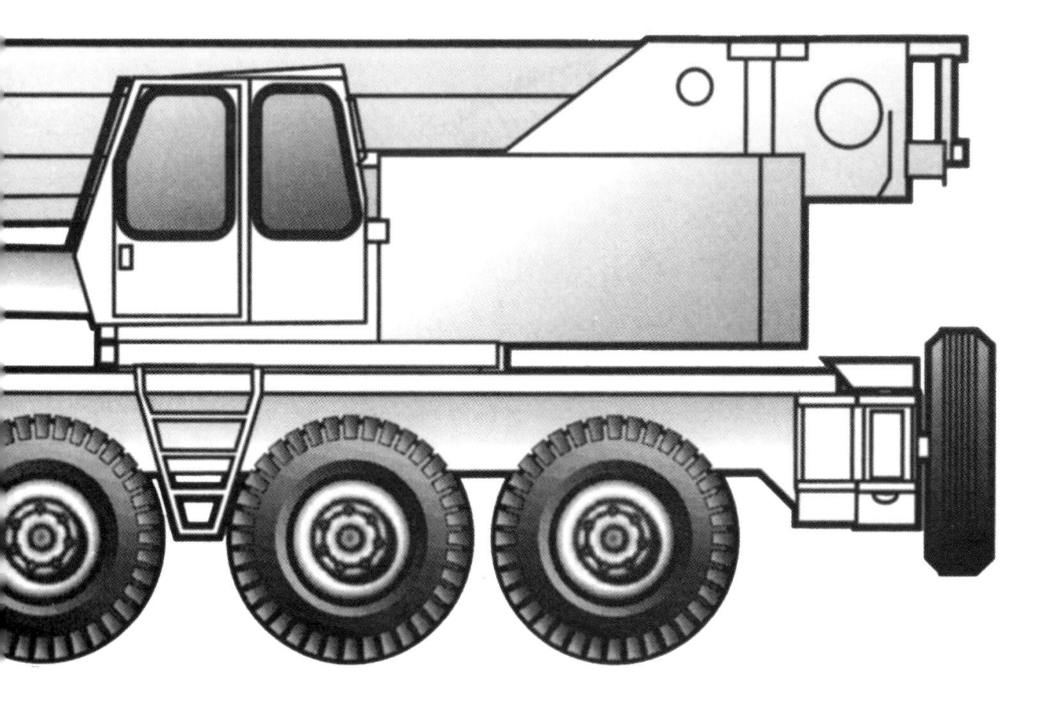

**MARCHETTI MG60.3 HYDRAULIC CRANE**

**Crane type**
All-terrain

**Boom length**
45.93–141.08ft | 14–43 m

**Rated lift capacity**
66.14t | 60 MT

**Engine**
Iveco 8460.41

**Power**
220 kW (300 hp) @ 2200 rpm

**Drive**
6x4

**Transmission**
10F, 1R

**Maximum speed**
40.89mph | 65.8 kph

**Tire size**
14.00R24

**Marchetti MG80.4 Hydraulic Crane**

This crane is powered by a V6 intercooler diesel and its overall dimensions are kept to a minimum to enable it to work in congested areas. It is 98.11 inches (2,490 millimeters) wide when traveling and 275.80 inches (7,000 millimeters) wide when its stabilizers are extended to the maximum. The overall length of the machine when traveling, including the boom, is 549.63 inches (13,950 millimeters). The Marchetti MG110.5 hydraulic crane is a 121.25-ton (110-tonne) rated version of the machine, and has five axles in order to handle the larger loads. The 148.81-ton (135-tonne) MG197.135 is the largest all-terrain crane made by Marchetti.

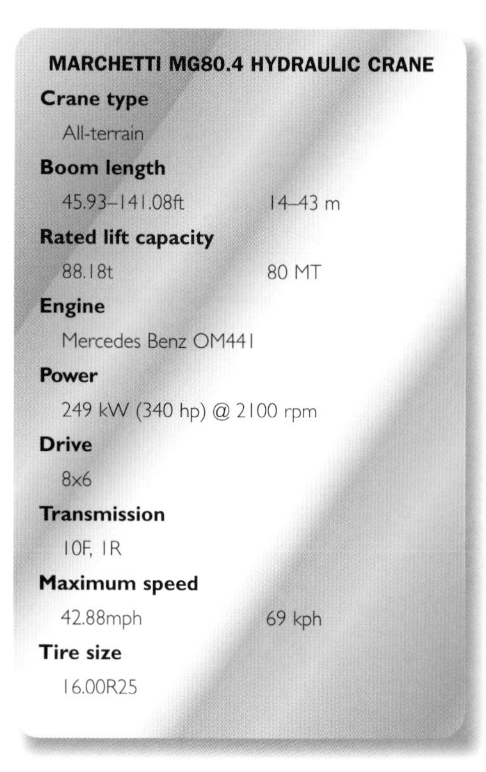

| MARCHETTI MG80.4 HYDRAULIC CRANE | |
|---|---|
| **Crane type** | |
| All-terrain | |
| **Boom length** | |
| 45.93–141.08ft | 14–43 m |
| **Rated lift capacity** | |
| 88.18t | 80 MT |
| **Engine** | |
| Mercedes Benz OM441 | |
| **Power** | |
| 249 kW (340 hp) @ 2100 rpm | |
| **Drive** | |
| 8x6 | |
| **Transmission** | |
| 10F, 1R | |
| **Maximum speed** | |
| 42.88mph | 69 kph |
| **Tire size** | |
| 16.00R25 | |

***The four axle Marchetti MG80.4 hydraulic crane has been designed to work in confined spaces and has two steering axles to enhance maneuverability.***

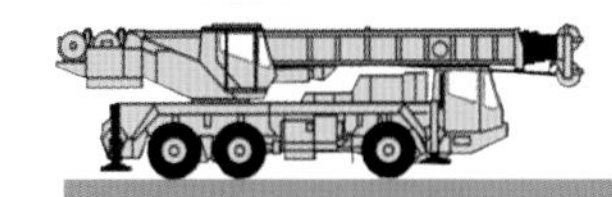



**PPM ATT290 All-terrain Crane**

The French company PPM manufacture a range of cranes with the designation of ATT. The Quadral family of eight machines ranges from the ATT240 to the ATT1190. The former has a maximum lifting capacity of 22.05 tons (20 tonnes), while the latter can lift up to 121.25 tons (110 tonnes). The ATT290 has a 30.86-ton (28-tonne) lifting capability, and a number of novel features including a tail swing that remains within the outrigger base, all-wheel steer and numerous options including a more powerful engine.

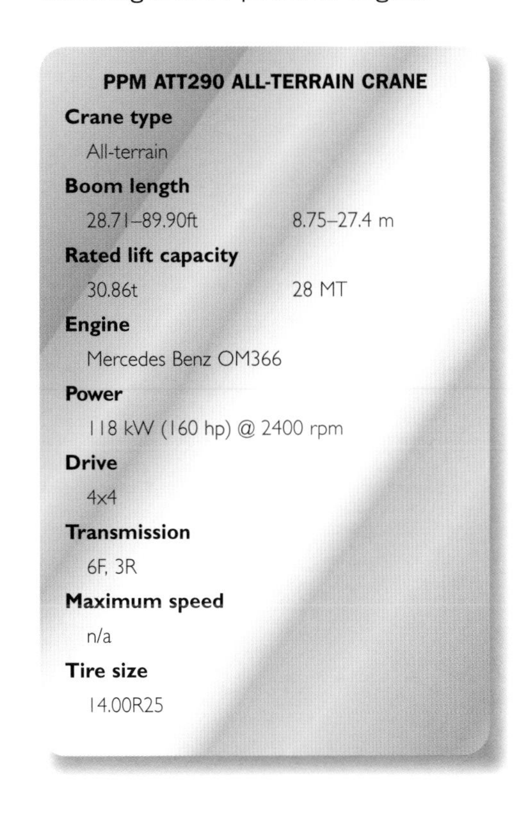

| PPM ATT290 ALL-TERRAIN CRANE | |
|---|---|
| **Crane type** | |
| All-terrain | |
| **Boom length** | |
| 28.71–89.90ft | 8.75–27.4 m |
| **Rated lift capacity** | |
| 30.86t | 28 MT |
| **Engine** | |
| Mercedes Benz OM366 | |
| **Power** | |
| 118 kW (160 hp) @ 2400 rpm | |
| **Drive** | |
| 4x4 | |
| **Transmission** | |
| 6F, 3R | |
| **Maximum speed** | |
| n/a | |
| **Tire size** | |
| 14.00R25 | |

***The designations of the PPM range of two axle cranes approximate to their lifting capabilities. The ATT 290 is one of the middleweights of the range.***

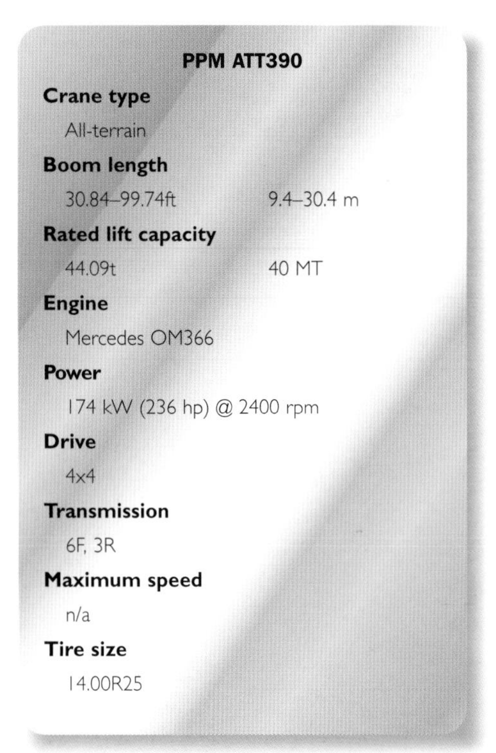

**PPM ATT390**

**Crane type**
All-terrain

**Boom length**
30.84–99.74ft 9.4–30.4 m

**Rated lift capacity**
44.09t 40 MT

**Engine**
Mercedes OM366

**Power**
174 kW (236 hp) @ 2400 rpm

**Drive**
4x4

**Transmission**
6F, 3R

**Maximum speed**
n/a

**Tire size**
14.00R25

***PPM's largest two axle hydraulic crane is the ATT390, it is powered by a Mercedes diesel engine.***

**PPM ATT590**

**Crane type**
All-terrain

**Boom length**
30.84–99.74ft 9.4–30.4 m

**Rated lift capacity**
65.04t 59 MT

**Engine**
Mercedes OM401

**Power**
203 kW (313 hp) @ 2400 rpm

**Drive**
6x6

**Transmission**
6F, 3R

**Maximum speed**
n/a

**Tire size**
14.00R25

***PPM's ATT 590 has three axles and a four section hydraulic boom.***

## PPM ATT390 & 590 All-terrain Cranes

These are larger capacity versions of the ATT290, based on 4x4 and 6x6 carriers and designed for use in difficult terrain. The key factors in these cranes performances are their hydraulic system, known as PPM Flowmatic, and self-levelling hydraulic suspension.

## PPM ATT1190

This 121.25-ton (110-tonne) capacity crane is the largest all-terrain machine manufactured by PPM. The carrier is powered by a V8, watercooled, turbo diesel engine, and unlike its smaller counterparts, the crane is powered by a separate engine, in this case a Mercedes OM 366 in-line, six-cylinder, turbo diesel. The boom is hydraulically operated and comprises five sections, which take 120 seconds to extend from minimum to maximum length.

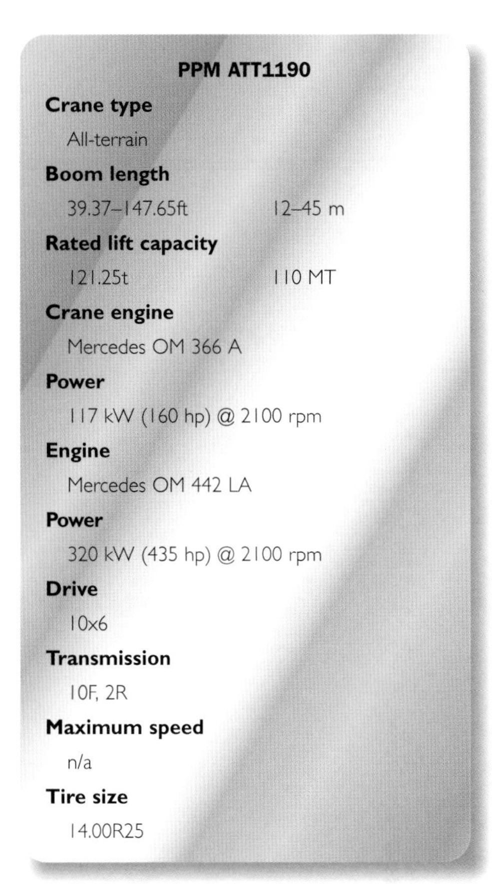

**PPM ATT1190**

**Crane type**
All-terrain

**Boom length**
39.37–147.65ft 12–45 m

**Rated lift capacity**
121.25t 110 MT

**Crane engine**
Mercedes OM 366 A

**Power**
117 kW (160 hp) @ 2100 rpm

**Engine**
Mercedes OM 442 LA

**Power**
320 kW (435 hp) @ 2100 rpm

**Drive**
10x6

**Transmission**
10F, 2R

**Maximum speed**
n/a

**Tire size**
14.00R25

***The ATT 1190 is a five axle crane and steers with all its axles to increase its usefulness in congested areas.***

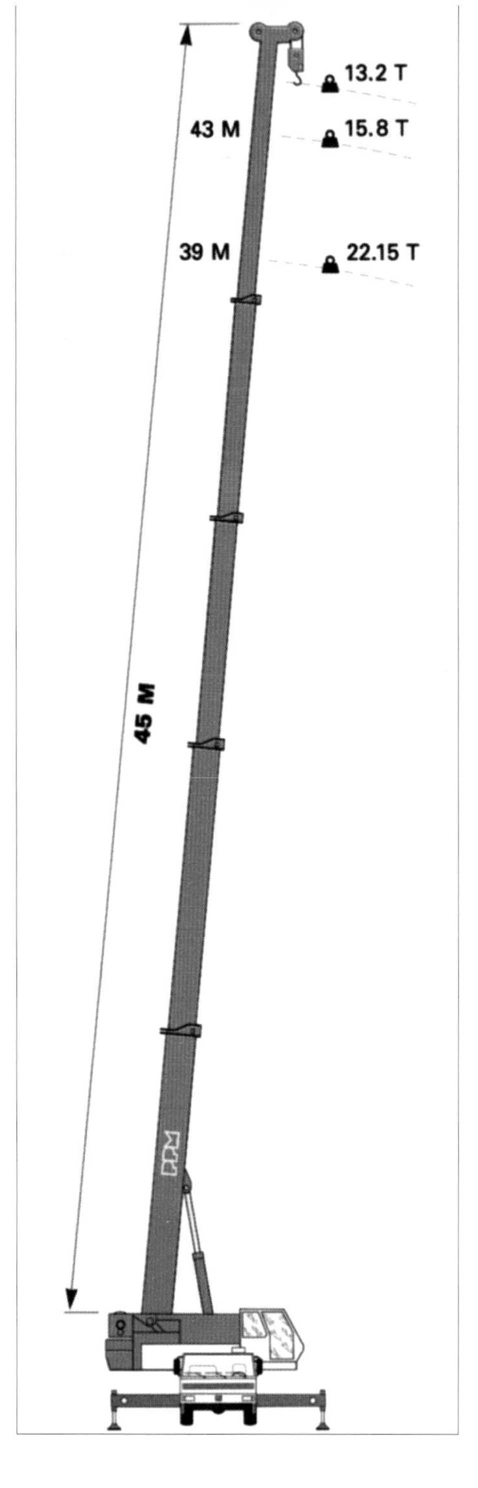


***The ATT 1190 has a maximum boom length of 147.65 feet (45 meters) and this diagram shows maximum lifts at various boom lengths.***

**Tadano Faun ATF120-5**

The ATF120-5 is the largest all-terrain crane in Tadano Faun's range of six machines. The smallest, the ATF35-2, has a maximum lifting capacity of 38.58 tons (35 tonnes). The five-axle 120-5 steers by turning the front three axles. Four axles are driven off-road to give a 10x8 drive pattern but of these five axles only three are driven when the crane is being used on surfaced roads.

| TADANO FAUN ATF120-5 | |
|---|---|
| **Crane type** | |
| All-terrain | |
| **Boom length** | |
| 41.01–116.67ft | 12.5–49 m |
| **Rated lift capacity** | |
| 132.28t | 120 MT |
| **Crane engine** | |
| Mercedes OM 366 A | |
| **Power** | |
| 112 kW (153 hp) @ 2000 rpm | |
| **Engine** | |
| Mercedes OM 442 LA | |
| **Power** | |
| 370 kW (503 hp) @ 2100 rpm | |
| **Drive** | |
| 10x8 | |
| **Transmission** | |
| 16F, 2R | |
| **Maximum speed** | |
| n/a | |
| **Tire size** | |
| 16.00R25 | |

***The side view of the Tadano Faun ATF120-5, the largest all-terrain crane in the range, has four driven and three steering axles.***

# Rough Terrain Cranes

**Manufacturers of this type of crane intended for construction industry and arduous terrain use include Bendini, Grove, Kato, Link-Belt, Locatelli Atlas, Lorain, P&H, PPM, and Tadano Faun. Many of these are smaller cranes in the up to 110.23-ton (100-tonne) category.**

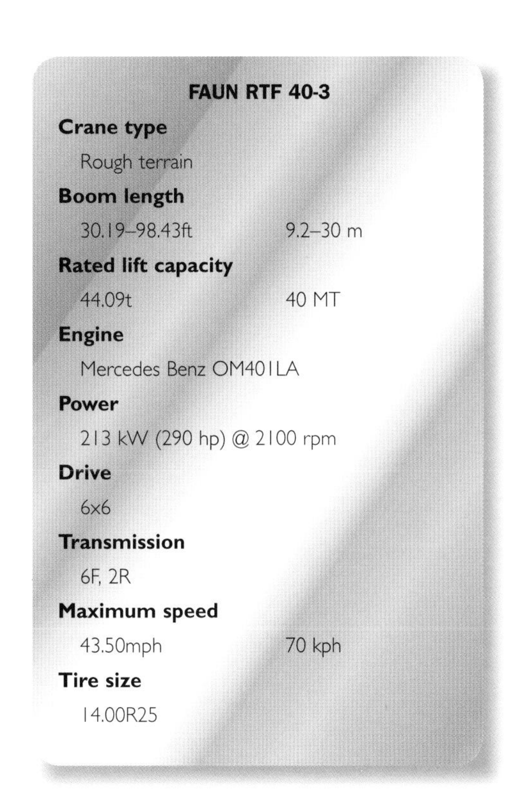

**FAUN RTF 40-3**

| | | |
|---|---|---|
| **Crane type** | | |
| | Rough terrain | |
| **Boom length** | | |
| | 30.19–98.43ft | 9.2–30 m |
| **Rated lift capacity** | | |
| | 44.09t | 40 MT |
| **Engine** | | |
| | Mercedes Benz OM401LA | |
| **Power** | | |
| | 213 kW (290 hp) @ 2100 rpm | |
| **Drive** | | |
| | 6x6 | |
| **Transmission** | | |
| | 6F, 2R | |
| **Maximum speed** | | |
| | 43.50mph | 70 kph |
| **Tire size** | | |
| | 14.00R25 | |

**Faun RTF 40-3 Rough Terrain Crane**

Faun, as part of Tadano Faun, produce the RTF40-3, a three-axle, rough terrain, hydraulic crane. It is designed for arduous ground conditions and has approach and departure angles of 16.5 degrees and 15 degrees respectively, making it less likely to become stuck on uneven ground. In traveling mode the machine, including the boom, has an overall length of 363.27 inches (9,220 millimeters).

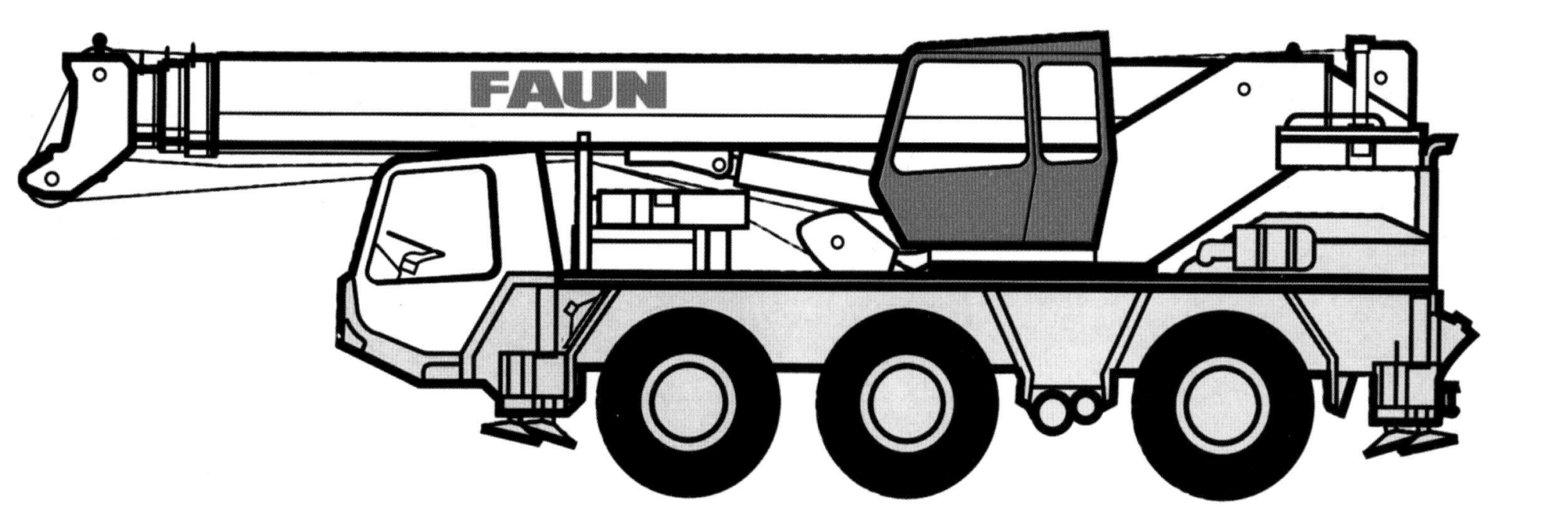


***A side elevation of the three axle Faun rough terrain crane, the RTF 40-3.***

***A Grove Cranes RT865B rough terrain crane in use on a construction site.***

***The Grove Cranes RT9100 is equipped with four wheel drive to enable it to access lifting operations in difficult terrain.***

**Grove RT9100 Hydraulic Crane**

The 99.21-ton (90-tonne) rated RT9100 is the largest rough terrain hydraulic crane made by Grove. Vehicles of this type feature four-wheel drive to enable them to get into position in difficult terrain, such as on large construction sites, before lowering their hydraulically operated outrigger jacks.

**GROVE RT865B HYDRAULIC CRANE**

**Crane type**
Rough terrain

**Boom length**
39.37–124.68ft 12–38 m

**Rated lift capacity**
66.14t 60 MT

**Engine**
Cummins 6CTA 8.3

**Power**
191 kW (250 hp) @ 2500 rpm

**Drive**
4x4

**Transmission**
6F, 6R

**Maximum speed**
24.86mph 40 kph

**Tire size**
33.25x29-26PR

**GROVE RT9100 HYDRAULIC CRANE**

**Crane type**
Rough terrain

**Boom length**
36.09–113.85ft 11–34.7 m

**Rated lift capacity**
99.21t 90 MT

**Engine**
Cummins 6CTA 8.3

**Power**
186 kW (250 hp) @ 2500 rpm

**Drive**
4x4

**Transmission**
6F, 6R

**Maximum speed**
16.16mph 26 kph

**Tire size**
33.25x35-32PR

***The Kato KR-300 ready for road travel with both boom and stabilisers retracted.***

| KATO KR-300 ROUGHTERR HYDRAULIC CRANE | |
|---|---|
| **Crane type** | |
| Rough terrain | |
| **Boom length** | |
| 29.20–93.18ft | 8.9–28.4 m |
| **Rated lift capacity** | |
| 33.07t | 30 MT |
| **Engine** | |
| Mitsubishi 6D16T | |
| **Power** | |
| 158 kW (221 hp) @ 2800 rpm | |
| **Drive** | |
| 4x4 | |
| **Transmission** | |
| 6F, 6R | |
| **Maximum speed** | |
| 35.42mph | 57 kph |
| **Tire size** | |
| 16x25-28PR | |

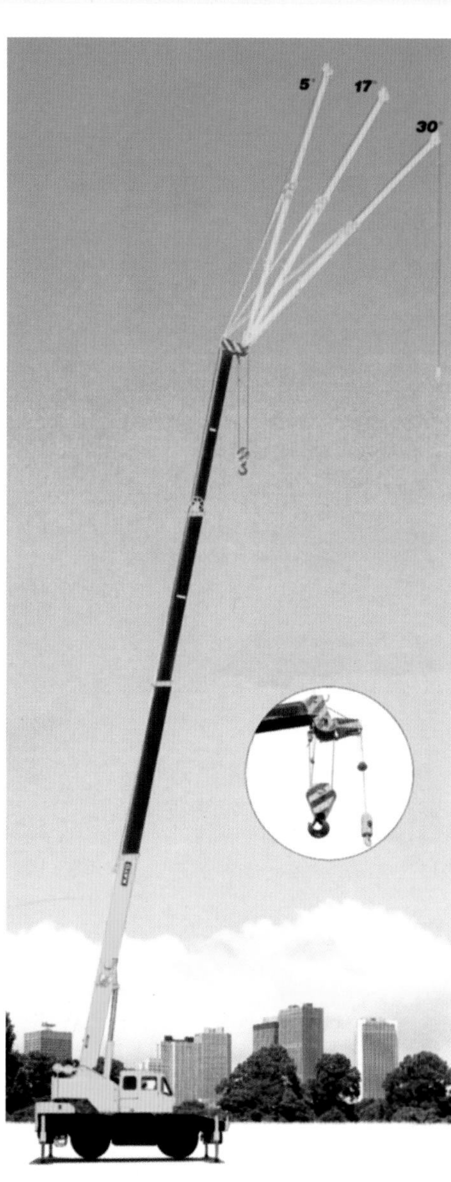


***The Kato KR-300 rough terrain crane has a hydraulic boom and an additional jib with three offsets for close in and high lift operation.***

**Kato KR-300 Hydraulic Crane**

The KR-300 has a 93.18-foot (28.4-meter) boom and a 39.70-foot (12.1-meter) jib which together provide a maximum lifting height of 54.28 yards (41.5 meters). The forward-acting derricking cylinder allows the derrick to be positioned from 0 degrees to 80 degrees, and the jib may be set at 5, 17, or 30 degrees which makes it useful in congested areas and confined spaces and allows items to be lifted up and over other buildings.

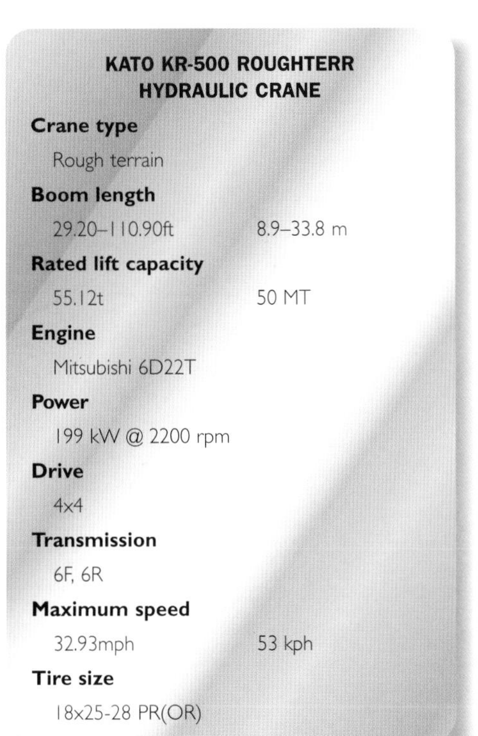

**KATO KR-500 ROUGHTERR HYDRAULIC CRANE**

**Crane type**
Rough terrain

**Boom length**
29.20–110.90ft    8.9–33.8 m

**Rated lift capacity**
55.12t    50 MT

**Engine**
Mitsubishi 6D22T

**Power**
199 kW @ 2200 rpm

**Drive**
4x4

**Transmission**
6F, 6R

**Maximum speed**
32.93mph    53 kph

**Tire size**
18x25-28 PR(OR)

**Kato KR-500 Hydraulic Crane**

The KR-500 is the largest of a range of three rough terrain cranes manufactured by the Japanese company Kato Works Co. Ltd., a long-established engineering company over a century old. The derricking range is from 0 degrees to 82 degrees and, like the KR-300, the fly jib may be set in one of three positions. A 62.34-foot (19-meter) luffing jib can be set in infinitely variable positions between 10 and 60 degrees.

***The Kato KR-500 rough terrain crane seen here ready to start lifting, has approach and departure angles of 22 degrees and 18 degrees respectively.***

# Crawler Cranes

| LIEBHERR LR11200 | |
|---|---|
| **Crane type** | |
| Crawler crane | |
| **Boom length** | |
| 137.80–321.54ft | 42–98 m |
| **Rated lift capacity** | |
| 440.92t | 400 MT |
| **Engine** | |
| Cummins KTA 38 C 1050 | |
| **Power** | |
| 746 kW (1,014 hp) @ 1800 rpm | |
| **Drive** | |
| Hydraulic | |
| **Transmission** | |
| Infinitely variable | |
| **Maximum speed** | |
| n/a | |
| **Track gauge** | |
| 41.01ft | 12.5 m |

**Manufacturers in this specialist field of crane construction include Grove, Liebherr, Link-Belt Construction Equipment, Manitowoc, Mannesmann Demag, P&H, PPM, RB, and Sumitomo. Koehring produce a specialist crawler-mounted material handler, the 6650 Skrapper. Grove Worldwide produce the HL150C, rated at 149.91-ton (136-tonne) lift capacity with a standard boom length of 49.87 feet (15.20 meters) and a maximum tip height of 177.51 yards (110.30 meters). They also produce special application cranes mounted on crawlers, such as the CM20. Link-Belt produces the lattice boom LS-718 and LS-818, with capacities of 250.22 and 299.83 tons (227 and 272 tonnes) respectively. They also produce a range of smaller models, both hydraulic crawler cranes and mechanical ones. The mechanical models are the 39.68 and 49.60-ton (36 and 45-tonne) LS-98D and LS-108D, while the hydraulic ones range from the 74.96-ton (68-tonne) LS-138H to the 200.07-ton (181.5-tonne) LS-248H.**

Liebherr's range of hydraulic crawler cranes consists of cranes with lifting capacities of between 38.58 and 132.28 tons (35 and 120 tonnes). The cranes are intended for use in civil engineering, bulk handling, aggregate extraction, and demolition, as well as draglining and clamshell grab operation. High line powers and speeds, winch pull ratings for high rope capacities per layer, combined with rugged structural design, mean the machines are reliable. The Litronic control system, standard on crawler models, combines electronics with hydraulic power transmission to make the carrier unit versatile.

**Liebherr LR11200 Crawler Crane**

The frame of this 440.92-ton (400-tonne) machine is manufactured by Liebherr and equipped with crawler carriers. Its traveling gear is maintenance-free tracklaying gear with flat track pads of 8.20-foot (2.5 meter) width. The drive includes two hydraulic traveling drives, with planetary gears, per crawler carrier. The crawler chains can be controlled independently and opposed so either direction can be forward or reverse.

The crane superstructure is connected to the crawler chassis by a slewing ring, the counterweight is 440.92 tons (400 tonnes), made up from a number of weights. The engine is a 12-cylinder diesel. At a 321.54-foot (98-meter) boom length, the crane can lift 545.64 tons (495 tonnes) and at a 137.80-foot (42-meter) boom length, 1,036.16 tons (940 tonnes).

# Mannesmann Demag Crawler Cranes

**Mannesmann Demag Baumaschinen, the construction equipment arm of the Mannesmann Demag group, build crawler-mounted cranes that offer high lifting capacities combined with low service weight. They offer high lifts and long outreach radii, a variety of boom permutations, and are noted as being reliable. Mannesmann Demag have developed a modular boom system that involves a variety of interchangeable components in order to increase the versatility of their products. As a result, high-load cranes can be adapted to a variety of lifting applications; the systems are known as Ringlift, Light Superlift, Heavy Superlift, and standard main boom.**

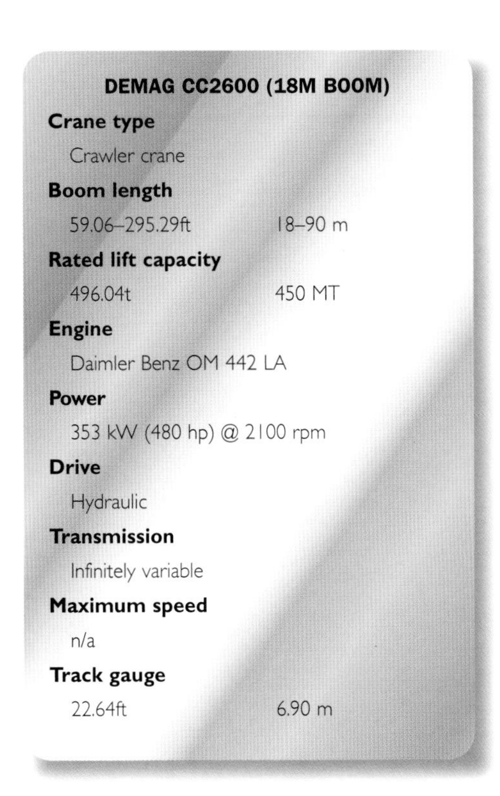

| DEMAG CC2600 (18M BOOM) | |
|---|---|
| **Crane type** | |
| Crawler crane | |
| **Boom length** | |
| 59.06–295.29ft | 18–90 m |
| **Rated lift capacity** | |
| 496.04t | 450 MT |
| **Engine** | |
| Daimler Benz OM 442 LA | |
| **Power** | |
| 353 kW (480 hp) @ 2100 rpm | |
| **Drive** | |
| Hydraulic | |
| **Transmission** | |
| Infinitely variable | |
| **Maximum speed** | |
| n/a | |
| **Track gauge** | |
| 22.64ft | 6.90 m |

***The Mannesmann Demag CC2600 rigged with two pairs of counterweights to balance the load to be lifted.***

*The Superlifter Demag CC2600 is available for use with a variety of boom combinations to give a variety of lifting ranges and capabilities.*

| DEMAG CC4800 (SH BOOM) | |
|---|---|
| **Crane type** | |
| Crawler crane | |
| **Boom length** | |
| 98.43–255.92ft | 30–78 m |
| **Rated lift capacity** | |
| 551.15t | 500 MT |
| **Engine** | |
| Mercedes Benz OM 424 LA | |
| **Power** | |
| 452 kW (615 hp) @ 2300 rpm | |
| **Drive** | |
| Hydraulic | |
| **Transmission** | |
| Infinitely variable | |
| **Maximum speed** | |
| 0.83mph | 1.34 kph |
| **Track gauge** | |
| 34.45ft | 10.50 m |

**Demag CC2600 (18m boom)**

The counterweight of 164.24 tons (149 tonnes) is carried separately on a carrier, and eight different boom combinations exist for various applications. The crawler tracks are mounted on side frames.

**Demag CC4800 (SH boom)**

Eight different boom combinations exist for various applications. The crawler tracks are mounted on side frames which are demountable for transportation. The crawlers are 47.57 feet (14.5 meters) long.

***A Mannesmann Demag crawler crane rigged with a complex arrangement of jibs and a counterweight.***

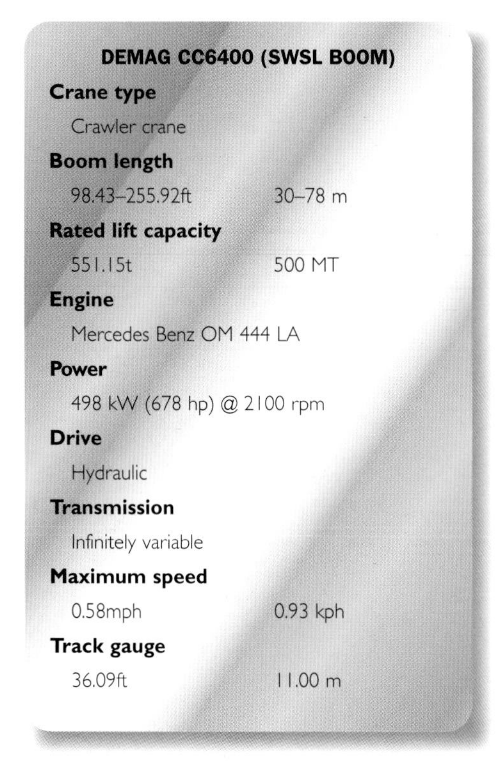

**DEMAG CC6400 (SWSL BOOM)**

| | | |
|---|---|---|
| **Crane type** | | |
| | Crawler crane | |
| **Boom length** | | |
| | 98.43–255.92ft | 30–78 m |
| **Rated lift capacity** | | |
| | 551.15t | 500 MT |
| **Engine** | | |
| | Mercedes Benz OM 444 LA | |
| **Power** | | |
| | 498 kW (678 hp) @ 2100 rpm | |
| **Drive** | | |
| | Hydraulic | |
| **Transmission** | | |
| | Infinitely variable | |
| **Maximum speed** | | |
| | 0.58mph | 0.93 kph |
| **Track gauge** | | |
| | 36.09ft | 11.00 m |

***The CC6400 in operation in an industrial lifting application.***

**Demag CC6400 (SWSL boom)**

Ten different boom combinations exist for various applications. The crawler tracks are mounted on side frames which are demountable for transportation. The crawlers are 48.89 feet (14.9 meters) long. The main boom is the same as used in smaller applications, but lengthened with inserts. In this application a 137.79–413.36-ton (125–375-tonne) counterweight is required.

**The twin lattice booms of the CC6400 while lifting.**

***The CC12000 in operation; it has a total of four lattice boom combinations to ensure its versatility in the heavy construction industry.***

| DEMAG CC12000 | |
|---|---|
| **Crane type** | |
| Crawler crane | |
| **Boom length** | |
| 177.17–374.03ft | 54–114 m |
| **Rated lift capacity** | |
| 1,047.19t | 950 MT |
| **Engine** | |
| Cummins KT A 38 C-1050 | |
| **Power** | |
| 783 kW (1,065 hp) @ 2100 rpm | |
| **Drive** | |
| Hydraulic | |
| **Transmission** | |
| Infinitely variable | |
| **Maximum speed** | |
| 0.30mph | 0.48 kph |
| **Track gauge** | |
| 40.03ft | 12.20 m |

### Demag CC12000

This Superlifter has a total weight of 1,554.24 tons (1,410 tonnes), including the counterweight, a 177.17-foot (54-meter) SH/LH boom, and hook block. Four lattice boom combinations are available to ensure the crane is capable of meeting the lifting requirements of heavy construction.

| SUPERLIFTER DEMAG CC12600 | |
|---|---|
| **Crane type** | |
| Crawler crane | |
| **Boom length** | |
| 177.17–767.75ft | 54–234 m |
| **Rated lift capacity** | |
| 1,047.19t | 950 MT |
| **Engine** | |
| Cummins KT A 38 C-1050 | |
| **Power** | |
| 1215 kW | |
| **Drive** | |
| Hydraulic | |
| **Transmission** | |
| Infinitely variable | |
| **Maximum speed** | |
| 0.30mph | 0.48 kph |
| **Track gauge** | |
| 32.15ft | 9.80 m |

### Superlifter Demag CC12600

This is the world's largest crawler crane for which four boom combinations are available. These give lifting ranges of 177.17–374.03 feet (54–114 meters) and 314.98–452.78 feet (96–138 meters) in SSL and LSL configurations, and of 177.17 + 137.80/374.03 + 275.60 feet (54 + 42 /114 + 84 meters) in SWSL and SFSL configurations. The machine weighs 66.14 tons (60 tonnes) and has within its equipment 4.23 miles (6.8 kilometers) of rope. It is demountable for transportation and its maximum transport width is a mere 13.12 feet (4 meters) and transport height 10.99 feet (3.35 meters).

| RB INTERNATIONAL PLC CH100 | |
|---|---|
| **Crane type** | |
| Crawler crane | |
| **Boom length** | |
| 59.06–236.23ft | 18–72 m |
| **Rated lift capacity** | |
| 110.23t | 100 MT |
| **Engine** | |
| Cummins LTA 10 | |
| **Power** | |
| 205 kW (275 hp) @ 1800 rpm | |
| **Drive** | |
| Hydraulic | |
| **Transmission** | |
| Infinitely variable | |
| **Maximum speed** | |
| n/a | |
| **Track gauge** | |
| 16.90ft | 5.15 m |

### RB International plc CH100 Hydraulic Crawler Crane

For ease of operation the CH100 is controlled with only two levers in a system which may be programed to suit the operator, who sits in a cab insulated against noise and vibration. The functions of the machine are displayed and monitored by a series of visual displays within the cab. The crawler tracks are deliberately long in order to provide a stable platform for all the boom lengths offered as options for this machine. The machine is also suited for dragline-type bucket operation.

***The Demag Superlifter CC12600, the world's largest crawler crane towering over a town where construction is about to commence.***

# On/Off-Highway Trucks

*The GMC 'deuce and a half' was a 6x6 truck built in massive numbers for the United States army of World War Two. Six wheel drive gave it both on- and off-road capability.*

# On/Off Highway Trucks

**Dual-purpose trucks have been under construction since the time of the First World War in both the United States and Europe. The rate of development meant that by the time of the Second World War the Allies had numerous all-wheel drive trucks. Notable American ones included trucks from Mack and GMC. The latter built the famous "Deuce and a half" 6x6 2.5 tonner (2.7 tonnes) that saw service in every theater of operations in huge numbers (it was also the basis of an altogether different type of off-highway truck, the DUKW amphibian). In Britain famous all-wheel drive trucks included the AEC Matador, and in the immediate postwar years, the Scammell Explorer. The Explorer was built for military use although it was one of a range that included models named Contractor and Mountaineer which were used in heavy construction and oil field development work.**

| MERCEDES BENZ UNIMOG | |
|---|---|
| **Model** | |
| U 2450L | |
| **Wheelbase** | |
| 181.24in | 4,600 mm |
| **Engine** | |
| Mercedes Benz OM 366 LA | |
| **Power** | |
| 177 kW @ 2600 rpm | |
| **Transmission** | |
| 8F, 2R | |
| **Drive** | |
| 6x6 | |
| **Tire size** | |
| 365/80R20 | |
| **Gross vehicle weight** | |
| 37,485lb | 17,000 kg |

# Mercedes Benz Unimog

***The Mercedes Benz Unimog is manufactured in both 4x4 and 6x6 variants for a variety of industrial applications where off-road ability is important.***

**Mercedes Benz of Germany have produced the Unimog for several decades. Although they are some of the most capable off-highway trucks ever manufactured, their small sizes precludes the majority of the range from inclusion here. The 6x6 U2450L model is one of the largest with a permissible gross vehicle weight of 37,485 pounds (17,000 kilograms). The Unimog features portal axles with disc brakes, an eight-speed manual transmission, and power steering.**

The Mercedes Benz Unimog is manufactured in numerous variants with specialist equipment and bodies for use in construction, road maintenance, industrial applications, fire fighting, and agricultural applications. It is manufactured in light and heavy duty applications and a variety of wheelbases, as well as six- and four-wheeled types.

One Unimog variant particularly relevant to this book is the ambulance bodied machine, a U 1550 L, that operates on the 3.86 square mile (10 square kilometer) area of the Hambach, Germany, open cast lignite mine operated by Rheinbraun AG. The medical side of the machine is fitted into the box body while the machine's off-highway prowess means that an accident can be reached regardless of the weather and terrain and tracks left by rigid dump trucks. Rheinbraun AG have used as many as 600 Unimogs since 1968 mainly as cargo and worker transport as well as mobile welding, lubrication, and workshop platforms.

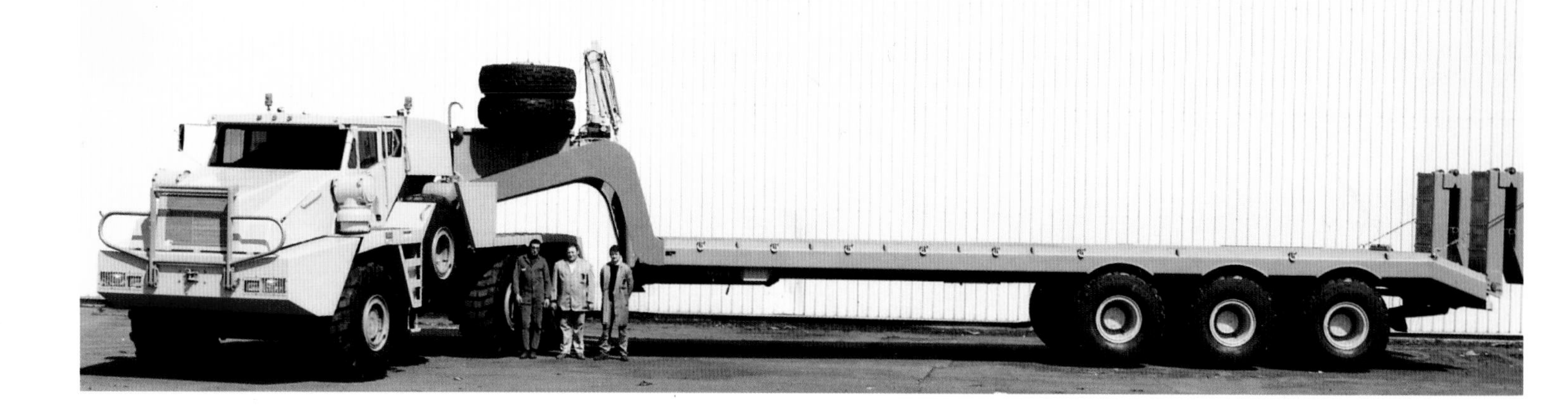


# MOL

***The MOL TB 800 Desert Lion all-terrain tractor with a semi-low loader semi-trailer.***

**MOL is a Belgian company founded by Gerard Mol in 1944, when he started making agricultural machinery. He gradually shifted into the production of trailers and semi trailers. MOL became a limited liability company in 1975. They manufacture a range of products, including a number of trucks designed for specialist on- and off-highway applications.**

**MOL All-terrain Tractor TB 800**

This machine is designed to tow a semitrailer through a fifth wheel type coupling. It features power steering, three rigid driven axles, hydropneumatic suspension, and a mounted cab. It is used in connection with an all-terrain trailer fitted with compatible wheels and tires.

**MOL ALL-TERRAIN TRACTOR TB 800**

**Model**
TB800

**Wheelbase**
n/a

**Engine**
MWM TBD 234 V12

**Power**
600 kW (816 hp) @ 2300 rpm

**Transmission**
Renk HS 227.18. 7F, 2R

**Drive**
6x6

**Tire size**
26.5R25

**Gross vehicle weight**
70,560lb 32,000 kg

***The MOL TB 800 has three driven axles to enhance its off-highway performance.***

***The MOL T6066 is designed for heavy transport in on- and off-road situations.***

**MOL Tractor T6066**

This specialist machine is designed to tow a low-bed semitrailer through a fifth wheel type coupling. It is intended for use for on- and off-road heavy transport. It features power steering, three rigid driven axles, hydropneumatic suspension, and a resiliently mounted cab. It is used in connection with an all-terrain trailer fitted with compatible wheels and tires.

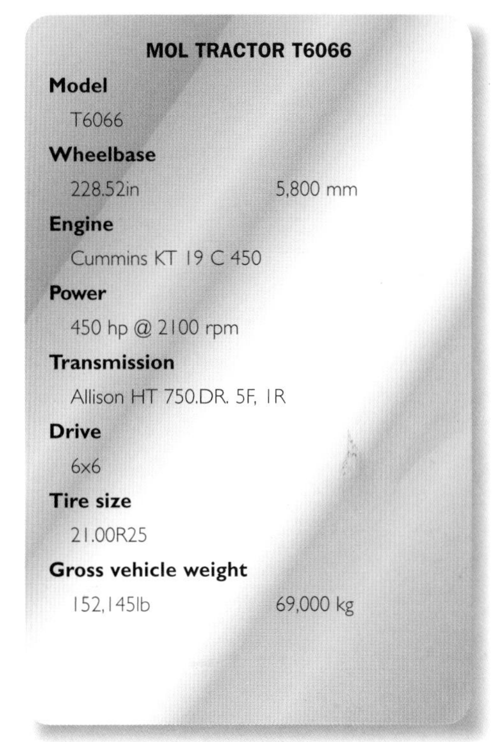

**MOL TRACTOR T6066**

**Model**
T6066

**Wheelbase**
228.52in 5,800 mm

**Engine**
Cummins KT 19 C 450

**Power**
450 hp @ 2100 rpm

**Transmission**
Allison HT 750.DR. 5F, 1R

**Drive**
6x6

**Tire size**
21.00R25

**Gross vehicle weight**
152,145lb 69,000 kg

***The T6066 is 6x6 and tows a semi-trailer connected by a fifth wheel coupling.***

**MOL Truck F6166**

This is a three-axle rigid truck designed for use in difficult terrain and climatic conditions. It features a steel cabover design of cab, mounted on a steel chassis, along with three driven axles made by Clark, a Clark gearbox, and a Caterpillar engine. The off-highway payload of the F6166 is 38.58 tons (35 tonnes).

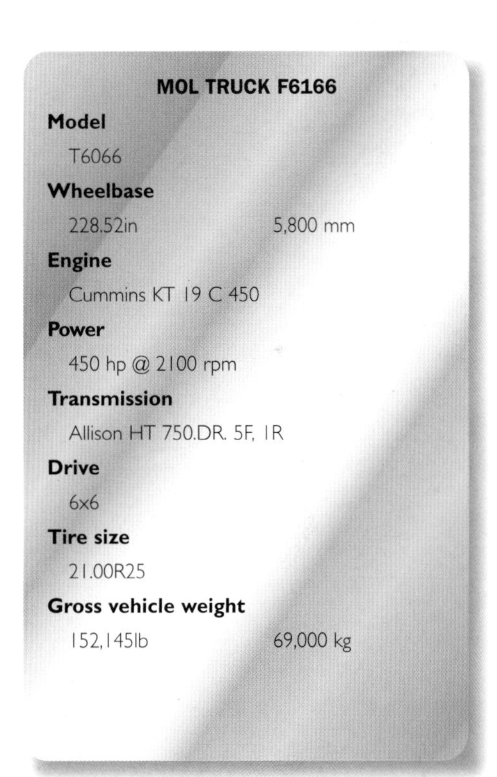

**MOL TRUCK F6166**

**Model**
T6066

**Wheelbase**
228.52in 5,800 mm

**Engine**
Cummins KT 19 C 450

**Power**
450 hp @ 2100 rpm

**Transmission**
Allison HT 750.DR. 5F, 1R

**Drive**
6x6

**Tire size**
21.00R25

**Gross vehicle weight**
152,145lb 69,000 kg

***The MOL F6166 is a three axle rigid truck with all axles driven.***

# Oshkosh Truck Corporation

***A truck from the Oshkosh S-Series.***

**The Oshkosh Truck Corporation was founded in 1917 and has always concentrated on trucks for arduous conditions. They were amongst the pioneers of the four wheel-drive concept for trucks. They have remained independent and based in Oshkosh, Wisconsin ever since the earliest days when the citizens of Oshkosh raised money for the development of the 4x4 truck by subscription. The Model A was their first machine and, like their current models, it relied on an engine bought from an engine manufacturer.**

### Oshkosh FF-Series

The FF-Series truck seen here, equipped with a rear-discharge cement mixer, is designed for severe duty and has all-wheel drive capability via an Oshkosh transfer case. It has power steering, driver controlled differential locks for the rear axles, a leaf sprung front axle, and Hendrickson rear suspension units. The whole machine is based around a steel frame and a conventional cab.

***The Oshkosh FF-Series truck equipped with a rear discharge cement mixer.***

| OSHKOSH FF-SERIES | |
|---|---|
| **Model** | |
| FF-115 | |
| **Wheelbase** | |
| 211.14in | 5,359 mm |
| **Engine** | |
| Cummins M11-330E+ | |
| **Power** | |
| 1,350lb/ft @ 1200 rpm | |
| **Transmission** | |
| Fuller RT-14609A 9 speed | |
| **Drive** | |
| 6x6 | |
| **Tire size** | |
| 12R22.5 | |
| **Gross vehicle weight** | |
| 61,012.35lb | 27,670 kg |

***The Oshkosh 8x6 S-Series forward placement concrete truck on site.***

**Oshkosh S-Series**

The Oshkosh S-Series forward placement concrete truck has the benefit of 8x6 wheel drive and is designed for off-highway operation. By delivering its load of concrete over the cab it is easier to operate in difficult terrain because of the greater visibility afforded from the cab. The mixer carries 11 cubic yards (8.41 cubic meters) of concrete. The S-Series truck has leaf sprung front suspension and Hendrickson rear. The chassis is made from steel and the cab is of the cabover type.

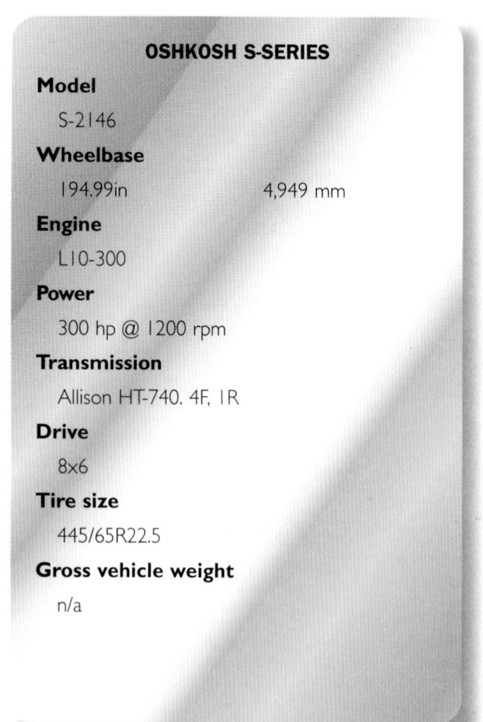

**OSHKOSH S-SERIES**

**Model**
S-2146

**Wheelbase**
194.99in 4,949 mm

**Engine**
L10-300

**Power**
300 hp @ 1200 rpm

**Transmission**
Allison HT-740. 4F, 1R

**Drive**
8x6

**Tire size**
445/65R22.5

**Gross vehicle weight**
n/a

A 6x6 snowplow variant of the Oshkosh P-Series truck designed with a heavy duty chassis for such tasks.

**Oshkosh P-Series Trucks**

Oshkosh manufacture two variants of the P-Series truck, four- and six-wheel drive. The wheelbase varies from 158.11 inches (4,013 millimeters) for the 4x4 to 175.13 inches (4,445 millimeters) for the 6x6. The trucks are designed as the bases of snow-removal trucks, and have an extra heavy-duty chassis to withstand the frontal and side loads imposed by snow clearing. The components used in the construction of the P-Series trucks include Dana axles, Allison transmissions, and a choice of either Detroit Diesel or Cummins engines. The cab is of the conventional design.

| OSHKOSH P-SERIES | |
|---|---|
| **Model** | |
| P-2523 | |
| **Wheelbase** | |
| 158.11in | 4,013 mm |
| **Engine** | |
| Detroit Series 60 | |
| **Power** | |
| 330 hp @ 2100 rpm | |
| **Transmission** | |
| Allison HT-740. 4F, 1R | |
| **Drive** | |
| 4x4 | |
| **Tire size** | |
| 395/85R20 | |
| **Gross vehicle weight** | |
| 48,009.47lb | 21,773 kg |

***The innovative Oshkosh M-1074 PLS-Palletised Load System – designed with military applications in mind.***

## Oshkosh PLS

The Palletized Load System was developed for the military as a new concept in logistics. The truck and trailer each carry a demountable cargo bed, called a flatrack capable of an 18.48-ton (16.77-tonne) load, giving a capacity of 36.96 tons (33.53 tonnes). The PLS has a hydraulically powered lifting hook in order to load and unload the PLS. A single pallet can be unloaded, or the entire truck and trailer can be unloaded, in less than five minutes. The all-wheel drive truck has the ability to cross rugged terrain easily in order to keep up with the speed and mobility of modern warfare.

**OSHKOSH PLS**

**Model**
M-1074

**Wheelbase**
223.87in — 5,682 mm

**Engine**
Detroit Diesel 8V92TA

**Power**
373 kW (500 hp) @ 2100 rpm

**Transmission**
Allison CLT-755 ATEC. 5F, 1R

**Drive**
10x10

**Tire size**
16.00R20

**Gross vehicle weight**
88,014.78lb — 39,916 kg

***The 4x4 Tiger TMV 40, the 6x6 variant is the TMV60. Its chassis cab design allows specialist rear bodies to be installed.***

# Tiger Engineering

**This company is based in Australia and produce various vehicles for the mining industry as well as carrying out conversions to dozers made by other companies. Many of the components used by Tiger are proprietary ones bought in from Caterpillar. Tiger have outlets in the United States and Canada, South America, and England.**

**Tiger TMV Multipurpose Vehicle**

This Australian-manufactured machine is designed for various applications in severe working environments. It is available in both 4x4 (TMV40) and 6x6 (TMV60) variants and brings together elements of on-highway trucks and mining equipment. Identical axles are used for parts commonality, and the machine is based on steel-box section unit. Power comes from a Caterpillar diesel engine, and a hydraulic suspension known as Hydroflex is used. The cab is of the forward-control type.

**TIGER TMV MULTIPURPOSE VEHICLE**

**Model**
TMV 40

**Wheelbase**
159.25in 4,042 mm

**Engine**
Caterpillar 3306

**Power**
373 kW (300 hp) @ 2200 rpm

**Transmission**
Caterpillar 5F, 1R

**Drive**
4x4

**Tire size**
26.50R25

**Gross vehicle weight**
132,300lb 60,000 kg

# List of Abbreviations

| | | | |
|---|---|---|---|
| in | inch(es) | mm | millimeter(s) |
| | | cm | centimeter(s) |
| ft | foot/feet | m | meter(s) |
| yd | yard(s) | | |
| $yd^3$ | cubic yard(s) | $m^3$ | cubic meter(s) |
| mph | miles per hour | kph | kilometers per hour |
| lb | pound(s) | kg | kilogram(s) |
| t | tons | MT | tonnes |
| PSI | pounds per square inch | kPa | kilograms per are (1 are = 100 square meters) |
| bhp | brake horsepower | | |
| hp | horsepower | | |
| kW | kilowatts | | |
| rpm | revolutions per minute | | |
| F | Forward (transmission) | | |
| R | Reverse (transmission) | | |

# Index

Page numbers in *italics* refer to picture captions

# Acknowledgments

The Publisher would like to thank the following companies for supplying material for publication:

Åkerman
Atlas
Aveling Barford
Bell Equipment
Bell South Africa
Bucyrus International Inc.
Cat Leverton
Daewoo
DDT Engineering
Euclid
Fiat-Hitachi
Finnings
Grove Worldwide
Hall Euro Enterprise Ltd
Heathfield Haulamatic Ltd
Inveco
JCB SCM Ltd
Kato Cranes (UK) Ltd
Kobelco
Komatsu Moxy
Krupp
Liebherr Great Britain
Liebherr Mining Truck Inc.
Mannesmann Demag
Marchetti
Mercedes Benz (United Kingdom) Ltd
O&K Orenstein & Koppel AG
Oshkosh Truck Corporation
PPM Grues Mobiles
RB International PLC
Samsung Heavy Industries Europe Ltd
Tadano Faun
Terex Equipment Ltd
Tiger
Volvo Construction Equipment Ltd

# Picture Credits

Chris Bacon pages 15, 16, 18, 29, 30, 31, 44, 49; John Carroll pages 67, 82 127, 146, 180; Ian Clegg pages 33, 64, 66, 77, 79, 86, 89, 90, 93, 114, 115, 147 a; R.H. Hooley page 12; Imperial War Museum pages 13, 92; MSC Co. pages 6,8,10; Peter Newark's American Pictures pages 11,14; Nigel Shuttleworth page 130.

ÅKERMAN
TIGER
ENGINEERING PTY LTD
TADANO
PPM
JCB
GROVE CRANE
VOLVO
EUCLID
CAT
Aveling Barford
HALLA
BELL EQUIPMENT
KATO
KOBELCO
LIEBHERR
KOMATSU
O&K
GROVE CRANE
SAMSUNG
RB

ÅKERMAN
TIGER
ENGINEERING PTY LTD
TADANO
PPM

JCB

GROVE CRANE
VOLVO
EUCLID
CAT
Aveling Barford
HALLA

BELL EQUIPMENT
KATO
KOBELCO
LIEBHERR
KOMATSU